Further Praise for *The Case Against Reality*

"[Donald] Hoffman's breezy fluency draws the reader with him. . . . [F]ascinating." —Jane O'Grady, *Spectator* (UK)

"A dense, lucid, and often unsettling exploration of how our brains interpret the world." —*Kirkus Reviews*

"A masterpiece of logic, rationality, science, and mathematics. Read this book carefully and you will forever change your understanding of reality, both that of the universe and your own self."

—Deepak Chopra, author of *The Healing Self*

"Hoffman's truly radical theory will force us to ponder reality in a completely different light. Handle with care. Your perception of the world around you is about to be dismantled!" —Chris Anderson, author of *TED Talks*

"A fresh view into who we truly are—one that transcends the perceptions that we accept as reality. Hoffman unapologetically takes us down a rabbit hole where we learn that all reality is virtual and that truth lies solely in you, the creator." —Rudolph Tanzi, coauthor of *Super Brain*

"Captivating and courageous . . . anyone who reads this book will likely never look at the world the same way again. Hoffman challenges us to rethink some of the most basic foundations of neuroscience and physics, which could prove to be exactly what we need to make progress on the most difficult questions we face about the nature of reality." —Annaka Harris, author of *Conscious*

"In the mood to have your mind blown? In this fascinating, deeply original, and wonderfully engaging book, Hoffman takes us on a tour of the uncharted territory where cognitive science, fundamental physics, and evolutionary biology meet—and where the nature of reality hangs in the balance. You'll never look at the world—or, rather, your interface—the same way again."

—Amanda Gefter, author of *Trespassing on Einstein's Lawn*

"Woody Allen once said, 'I hate reality, but . . . where else can you get a good steak dinner?' Hoffman turns that joke on its head: What we have always been after is the steak dinner; what we call reality is our best adapted strategy for getting it. Sink your teeth into that!"

—Christopher A. Fuchs, professor of physics,
University of Massachusetts Boston

"This book is a must-read if you want to bring your understanding of 'reality' in sync with the way the World is. You are in for some major surprises and mind expanding. A good read that will set you thinking about yourself, others, and the world." —Jan Koenderink, author of *Color for the Sciences*

The Case Against Reality

The Case Against Reality

Why Evolution Hid the Truth from Our Eyes

DONALD HOFFMAN

W. W. NORTON & COMPANY
Celebrating a Century of Independent Publishing

Printed in the United States of America
First published as a Norton paperback 2022

For information about permission to reproduce selections from this book, write to
Permissions, W. W. Norton & Company, Inc., 500 Fifth Avenue, New York, NY 10110

For information about special discounts for bulk purchases, please contact
W. W. Norton Special Sales at specialsales@wwnorton.com or 800-233-4830

Manufacturing by Lakeside Book Company
Book design by Daniel Lagin
Production manager: Lauren Abbate

Library of Congress Cataloging-in-Publication Data

Names: Hoffman, Donald D., author.
Title: The case aganist reality : why evolution hid the truth from our eyes / Donald Hoffman.
Description: First Edition. | New York : W.W. Norton & Company, 2019. | Includes bibliographical references and index.
Identifiers: LCCN 2019006962 | ISBN 9780393254693 (hardcover)
Subjects: LCSH: Reality. | Perception (Philosophy)
Classification: LCC BD331 .H567 2019 | DDC 121/.34—dc23
LC record available at https://lccn.loc.gov/2019006962

ISBN 978-0-393-54148-9 pbk.

W. W. Norton & Company, Inc., 500 Fifth Avenue, New York, N.Y. 10110
www.wwnorton.com

W. W. Norton & Company Ltd., 15 Carlisle Street, London W1D 3BS

1 2 3 4 5 6 7 8 9 0

To Joaquin, Noemi, and Cayetano,
I offer the red pill.

I think that tastes, odors, colors, and so on . . .
reside in consciousness. Hence if the living
creature were removed, all these qualities
would be wiped away and annihilated.

—GALILEO GALILEI

CONTENTS

PREFACE

Your eyes will save your life today. With their guidance, you will not tumble down stairs, leap before a speeding Maserati, grab the tail of a rattlesnake, or munch on a moldy apple.

Why are our eyes, and all of our senses, reliable guides? Most of us have a hunch: they tell us the truth. The real world, we assume, consists of cars and stairs and other objects in space and time. They exist even if no living creature observes them. Our senses are simply a window on this objective reality. Our senses do not, we assume, show us the whole truth of objective reality. Some objects are too small or too far away. On rare occasions our senses are even wrong—artists, psychologists, cinematographers, and others can cook up illusions that fool them. But normally our senses report the truths we need to navigate safely through life.

Why do our senses exist to reveal the truth? Again, we have a hunch: evolution. Those of our ancestors who saw reality more accurately had an advantage over those who saw it less accurately, especially in critical activities such as feeding, fighting, fleeing, and mating. As a result, they were more likely to pass on their genes, which coded for more accurate perceptions. We are the offspring of those who, in each generation, saw objective reality more accurately. Therefore, we can be confident that we see it accurately. Our hunch, in short, is that truer perceptions are fitter perceptions. Evolution weeds out untrue perceptions. That is why our perceptions are windows on objective reality.

These hunches are wrong. On the contrary, our perceptions of snakes and apples, and even of space and time, do not reveal objective reality. The problem is not that our perceptions are wrong about this or that detail. It's that the very language of objects in space and time is simply the wrong language to describe objective reality. This is not a hunch. It is a theorem of evolution by natural selection that wallops our hunches.

The idea that our perceptions mislead us about objective reality, in whole or in part, has a long history. Democritus, around 400 BCE, famously claimed that our perceptions of hot, cold, sweet, bitter, and color are conventions, not reality.[1] A few decades later, Plato likened our perceptions and conceptions to flickering shadows cast on the walls of a cave by an unseen reality.[2] Philosophers ever since have debated the relation between perception and reality. The theory of evolution injects new rigor into this debate.

How can our senses be useful—how can they keep us alive—if they don't tell us the truth about objective reality? A metaphor can help our intuitions. Suppose you're writing an email, and the icon for its file is blue, rectangular, and in the center of your desktop. Does this mean that the file itself is blue, rectangular, and in the center of your computer? Of course not. The color of the icon is not the color of the file. Files have no color. The shape and position of the icon are not the true shape and position of the file. In fact, the language of shape, position, and color cannot describe computer files.

The purpose of a desktop interface is not to show you the "truth" of the computer—where "truth," in this metaphor, refers to circuits, voltages, and layers of software. Rather, the purpose of an interface is to hide the "truth" and to show simple graphics that help you perform useful tasks such as crafting emails and editing photos. If you had to toggle voltages to craft an email, your friends would never hear from you.

That is what evolution has done. It has endowed us with senses that hide the truth and display the simple icons we need to survive long enough to raise offspring. Space, as you perceive it when you look around, is just your desktop—a 3D desktop. Apples, snakes, and other physical objects are simply icons in your 3D desktop. These icons are useful, in part, because they hide the complex truth about objective reality. Your senses have evolved to give you what you need. You may want truth, but you don't need truth. Perceiving truth would drive our species extinct. You need simple icons that show you how to act to stay alive. Perception is not a window on objective reality. It is an interface that hides objective reality behind a veil of helpful icons.

"But," you ask, "if that speeding Maserati is just an icon of your interface, why don't you leap in front of it? After you die, then we'll have proof that a car is not just an icon. It's real and it really can kill."

I wouldn't leap in front of a speeding car for the same reason I wouldn't carelessly drag my blue icon to the trashcan. Not because I take the icon literally—the file is not blue. But I do take it seriously: if I drag the icon to the trashcan, I could lose my work.

And that is the point. Evolution has shaped our senses to keep us alive. We have to take them seriously: if you see a speeding Maserati, don't leap in front of it; if you see a moldy apple, don't eat it. But it is a mistake of logic to assume that if we must take our senses seriously then we are required—or even entitled—to take them literally.

I take my perceptions seriously, but not literally. This book is about why you should do the same, and why that matters.

I explain why evolution hid objective reality and endowed us instead with an interface of objects in space and time. Together, we will explore how this counterintuitive idea dovetails with discoveries in physics that are equally counterintuitive. And we will examine how our interface works and how we manipulate it with makeup, marketing, and design.

In chapter one, we confront the greatest unsolved mystery in science: your experience of the taste of dark chocolate, the smell of crushed garlic, the blare of a trumpet, the sensual feel of plush velvet, the sight of a red apple. Neuroscientists have found many correlations between such conscious experiences and brain activity. They have discovered that our consciousness can be split in half with a scalpel, and the two halves can have different personalities, with different likes, dislikes, and religious beliefs: one-half can be an atheist while the other believes in God. But despite all this data, we still have no plausible story about how brain activity might generate a conscious experience. This stunning failure suggests that we have made a false assumption. Hunting for a culprit led me to look more closely at how our senses are shaped by natural selection.

A clear example of this shaping is our sense of beauty. We explore, in chapter two, beauty and attraction through the lens of evolution. When you glance at another person, you immediately—and unconsciously—pick up dozens of sensory clues, and run them through a sophisticated algorithm, forged by evolution, that decides one thing: reproductive potential—the likelihood that this person could successfully raise offspring. Your algorithm, in a fraction of a second, summarizes its complex analysis with a simple feeling—ranging from hot to not. Through the course of the chapter, we examine specific clues of beauty in the human eye. Men are attracted to women with larger eyes that have larger irises, larger pupils, slightly bluish scleras (the whites of the eyes), and distinctive limbal rings—the dark border between the iris and the sclera. What women want is more complex, and it's a fascinating story that we will examine more closely. As we survey our sense of beauty, we absorb key concepts of evolution, learn useful tricks to spiff up portraits, and explore the logic of natural selection—including the logic that tempts us to deceive others by spiffing up.

Many experts in evolution and neuroscience claim that our senses evolved to report truths about objective reality. Not the full spectrum of truth—just what we need to raise kids. We listen to these experts in chapter three. We hear from Francis Crick who discovered, along with James Watson, the structure of DNA. In a series of letters that Crick and I exchanged a decade before his death, he argues that our perceptions match reality, and that the sun existed before anyone saw it. We hear from David Marr, a professor at MIT who combined insights from neuroscience and artificial intelligence to transform the study of human vision. In his classic book *Vision*, Marr contends that we evolved to see a true description of objective reality.

Marr was my doctoral advisor until his death at age thirty-five; he influenced my early ideas, and those of the entire field, on this topic. We hear from Robert Trivers, an insightful evolutionary theorist who maintains that our senses evolved to give us an accurate view of reality. Philosophers have long wondered, "Can we trust our senses to tell us truths about reality?" Many brilliant scientists answer, "Yes."

We look, in chapter four, at the case for "No." We encounter a startling "Fitness-Beats-Truth" (FBT) theorem, which states that evolution by natural selection does not favor true perceptions—it routinely drives them to extinction. Instead, natural selection favors perceptions that hide the truth and guide useful action. Without equations or Greek symbols, we explore the new field of evolutionary game theory, which allows Darwin's ideas to be transformed into precise mathematics that lead to this shocking theorem. We look at computer simulations of evolutionary games, which confirm the predictions of the FBT Theorem. We find further confirmation from simulations of genetic algorithms, in which perceptions and actions coevolve.

The FBT Theorem tells us that the language of our perceptions—including space, time, shape, hue, saturation, brightness, texture, taste, sound, smell, and motion—cannot describe reality as it is when no one looks. It's not simply that this or that perception is wrong. It's that none of our perceptions, being couched in this language, could possibly be right.

At this point, our intuitions falter: How could our senses be useful if they don't report the truth? In chapter five, we aid our intuitions by exploring an interface metaphor. Space, time, and physical objects are not objective reality. They are simply the virtual world delivered by our senses to help us play the game of life.

"Well," you might say, "if you claim that space, time, and objects are not objective reality, then you are straying into the turf of physics, and physicists will be happy to set you straight." In chapter six, we discover that eminent physicists admit that space, time, and objects are not fundamental; they're rubbing their chins red trying to divine what might replace them. Some say that spacetime—a union of space and time required by Einstein's theories of relativity—is doomed.[3] They say that it is a hologram, made out of bits of information. Others say that reality differs from one observer to another, or that the history of the universe is not fixed but depends on what is observed now. Physics and evolution point to the same conclusion: spacetime and objects are not foundational. Something else is more fundamental, and spacetime emerges from it.

If spacetime is not a foundational, preexisting stage on which the drama of the universe unfolds, then what is it? In chapter seven, we wade into the curious and curiouser: spacetime is just a data format—much like data structures in your mobile device—that serves to keep us alive. Our senses report fitness, and an error in this report could ruin your life. So our senses use "error-correcting codes" to detect and correct errors. Spacetime is just a format our senses use to report fitness payoffs and to correct errors in these reports. To see how this works, we play with some visual illusions, and catch ourselves in the act of correcting errors. Then we use these insights to have fun with clothing: we can manipulate the visual codes to help men and women look even better in their jeans—by making careful alterations to stitches, pockets, finishes, and embroideries.

Then we look at color. From the refreshing blue of clear skies to the vibrant green of spring grasses, our rich world of light and color is a welcome gift, compliments of four kinds of photoreceptors in the eye. But *Arabidopsis thaliana*, a small weed that looks like wild mustard, has eleven kinds of photoreceptors.[4] The lowly cyanobacterium, which has colonized the earth for at least two billion years, boasts twenty-seven.[5] In chapter eight, we discover that color is a code for messages about fitness used by many species, a code that excels at compressing data much as you might compress a photo before texting it to a friend. Colors can trigger emotions and memories that enhance our fitness by guiding our actions. Corporations harness the power of color as a tool for branding, and will go to great lengths to defend a color as intellectual property. But as potent and evocative as color may be, "chromatures," which are textured colors, prove far more versatile and powerful than colors alone, and for good evolutionary reasons. Chromatures can be designed to trigger specific emotions and associations. If you understand our codes for fitness, then you can intelligently hack them for your benefit.

But evolution is not done with our sensory codes for fitness. It still experiments with novel interfaces for our enterprising species. Four percent of us are "synesthetes" who perceive a world that differs from the norm. We meet Michael Watson, who felt with his hands what he tasted with his mouth: when he tasted spearmint he felt tall, cold columns of glass; angostura bitters felt like "a scraggly basket of hanging ivy." Each taste had its own 3D object, which he created in the moment of taste and destroyed when he stopped tasting. Some synesthetes see a unique color for each number, letter, day of the week, or month of the year—and excel at discerning colors.

Perception may seem effortless, but in fact it requires considerable energy. Each precious calorie you burn on perception is a calorie you must find and take from its owner—perhaps a potato or an irate wildebeest. Calories can be difficult and dangerous to procure, so evolution has shaped our senses to be misers. One consequence, we discover in chapter nine, is that vision cuts corners: you see sharp detail only within a small circular window, whose radius is the width of your thumb held at arm's length. If you close one eye and hold out your thumb, you can see just how tiny it is. We think we see the whole field of vision in great detail, but we've been duped: each place we look falls into that small window of sharp detail, so we mistakenly assume that we see everything in detail. Only within that small window does your sensory interface construct a detailed report of fitness payoffs. That crucial report is formatted as the shape, color, texture, motion, and identity of a physical object. You create a suitable object—your description of payoffs—with a glance. You destroy it and create another with your next glance. Your wide field of vision guides your eyes to attend where there are vital payoffs to report, and thus an object to create. We explore the rules that govern attention, how they apply in marketing and design, and how an ad can, by accident, promote a rival if it flouts the rules.

If our senses hide reality behind an interface, then what is that reality? I don't know. But in chapter ten we explore the idea that conscious experiences are fundamental. When you look at yourself in a mirror you see skin, hair, eyes, lips, and the expression of your face. But you know that hidden behind your face is a far richer world: your dreams, fears, politics, love of music, taste in literature, love of family, and experiences of colors, smells, sounds, tastes, and touches. The face you see is just an interface. Behind it is the vibrant world of your experiences, choices, and actions.

Perhaps the universe itself is a massive social network of conscious agents that experience, decide, and act. If so, consciousness does not arise from matter; this is a big claim that we will explore in detail. Instead, matter and spacetime arise from consciousness—as a perceptual interface.

This book offers you the red pill.[6] If you can accept that the technology of virtual reality will one day create for you a compelling experience that is nothing like your experience when you take off the headset, then why be so certain that, when you remove the headset, you're seeing reality as it is? The purpose of this book is to help you take off the next headset, the one you didn't know you were wearing all along.

The Case Against Reality

CHAPTER ONE

Mystery

The Scalpel That Split Consciousness

"How it is that anything so remarkable as a state of consciousness comes about as a result of irritating nervous tissue, is just as unaccountable as the appearance of the Djinn, when Aladdin rubbed his lamp."

—THOMAS HUXLEY, *THE ELEMENTS OF PHYSIOLOGY AND HYGIENE*

"'A motion became a feeling!'—no phrase that our lips can frame is so devoid of apprehensible meaning."

—WILLIAM JAMES, *THE PRINCIPLES OF PSYCHOLOGY*

In February of 1962, Joseph Bogen and Philip Vogel sliced in half the brain of Bill Jenkins—intentionally, methodically, and with careful premeditation. Jenkins, then in his late forties, recovered and went on to enjoy a quality of life that had eluded him for years. In the decade that followed, Bogen and Vogel split brain after brain in California, earning them the epithet "the West Coast butchers."[1]

Each brain that they split belonged to a person who suffered from severe and intractable epilepsy, a condition caused by abnormal neural activity racing through the brain. The best drugs available at that time failed these epileptics, leaving them vulnerable to a seizure, a convulsion, or a "drop attack"—a sudden loss of muscle tone that often caused a damaging fall. Normal life evaded them: they couldn't drive, work, or enjoy a carefree night at a ball game. Daily existence devolved into drudgery, punctuated by episodes of horror.

Bogen and Vogel were talented neurosurgeons based at the University of Southern California and the California Institute of Technology. They split the brains of epileptics in a daring attempt to quarantine the anomalous neural activity that ravaged their lives.

The surgery was delicate and intricate, but its idea was simple. The human brain harbors 86 billion neurons that converse in an electrochemical dialect—a vast social network, each member following and being followed, as if they were tweeting and retweeting, each in its own unique style. Each neuron tweets via its axon and follows via its dendrites. This network, despite its complexity, is normally stable, allowing an orderly flow of messages. But just as a collision of cars can disrupt, in widening ripples, the flow of traffic in a city, so also a sudden surfeit of aberrant signals in the brain can disrupt the flow of electrochemical messages through the brain, triggering seizures, convulsions, and loss of consciousness.

Bogen and Vogel sought to halt the disastrous ripples before they swamped the brain. Fortunately, the anatomy of the brain itself suggests an opportune place and method. The brain is divided into two hemispheres, left and right. Each hemisphere has 43 billion neurons. Their axons subdivide, like branches of a tree, to allow trillions of links among them. But, in contrast to the rich interconnections within a hemisphere, the bond between hemispheres is a tiny cable, the corpus callosum, with just over 200 million axons—roughly one axon between hemispheres for every two hundred within a hemisphere. This bottleneck offers an ideal place to cut, and thereby to halt the spread of debilitating ripples from one hemisphere to the other. This scheme is admittedly crude, much like trying to stop the spread of a computer virus from Europe to the Americas by cutting all cables across the Atlantic. But triage was necessary. Bogen and Vogel chose to let one hemisphere endure the fury of epilepsy, in hopes that the other hemisphere, and thus the patient, might suffer less.

The surgery, known technically as a "corpus callosotomy" and informally as a "split-brain operation," was a clinical success. Bill Jenkins suffered no more drop attacks, and just two general convulsions in the next ten years. Other patients enjoyed similar relief. One attended a ball game in person for the first time in years, and another landed a full-time job for the first time in his life. Callosotomy was soon regarded not as "West Coast butchery" but as "a possible new treatment modality."

When I first met Bogen in 1995, our topic of discussion was not the dramatic success of his surgery, but the exotic changes in consciousness that it triggers. Joe had been invited to speak at a meeting of the Helmholtz Club, a small group of neuroscientists, cognitive scientists, and philosophers that, for many years, met monthly at UC Irvine. The purpose of the club was to explore how advances in neuroscience might spawn a scientific theory of consciousness. We met in Irvine because its central location was convenient for members as far north as Cal Tech, USC, and UCLA, and as far south as UC San Diego and the Salk Institute. We met in secret to avoid interlopers attracted by the fame of one club member, Francis Crick, who had focused his powerful intellect on the mystery of consciousness. We started our meetings with a buffet lunch at the University Club at UC Irvine, then spent the afternoon in a private room, grilling two invited speakers until six o'clock. We then retired to a restaurant, usually near South Coast Plaza, and continued deliberating late into the night.

The mystery of consciousness, which was the focus of the Helmholtz Club and the subject of Bogen's talk, is quite simply the mystery of who we are. Your body, like other objects, has physical attributes such as position, mass, and velocity. If, heaven forbid, a rock and your body fell simultaneously from the Leaning Tower of Pisa, both would strike the ground at the same time.

On the other hand, we differ from rocks in two key respects. First, we experience sensations. We taste chocolate, suffer headaches, smell garlic, hear trumpets, see tomatoes, feel dizzy, and enjoy orgasms. If rocks have orgasms, they're not letting on.

Second, we have "propositional attitudes," such as the belief that rocks don't have headaches, the fear that stocks might fall, the wish to vacation in Tahiti, and the wonder why Chris won't call. Such attitudes allow us to predict and interpret our behavior and that of others. If you wish to vacation in Tahiti and believe that you'll need an airline ticket to do so, then there's a good chance you'll buy that ticket. Your propositional attitudes predict and explain your behavior. If Chris calls and says he'll arrive on the train tomorrow morning at nine o'clock, then your attribution of propositional attitudes to Chris—that he wants and intends to take the train—allows you to predict where he will be tomorrow at nine, indeed with greater facility than if you knew the state of each particle of his body.

Like a rock, we have bona fide physical properties. But unlike a rock, we have conscious experiences and propositional attitudes. Are these also physical? If so, it's not obvious: What is the mass of dizziness, the velocity of a headache, or the position of the wonder why Chris won't call? In each case, the question itself seems to harbor confusion, and to mismatch categories. Dizziness is not the kind of thing that can be weighed on a scale; a wonder has no spatial coordinates; a headache can't be clocked with a radar gun.

But conscious experiences and propositional attitudes are essential to human nature. Delete them and we lose our very selves. The bodies that remained would lumber through life pointlessly.

So, what kind of creature are you? How is your body related to your conscious experiences and propositional attitudes? How is your experience of a chai latte related to activities in your brain? Are you just a biochemical machine? If so, how does your brain give rise to your conscious experiences? The question is deeply personal and, as it happens, deeply mysterious.

The German mathematician and philosopher Gottfried Leibniz grasped the mystery in 1714: "It must be confessed, however, that Perception, and that which depends upon it, are inexplicable by mechanical causes, that is to say, by figures and motions. Supposing that there were a machine whose structure produced thought, sensation, and perception, we could conceive of it as increased in size with the same proportions until one was able to enter into its interior, as he would into a mill. Now, on going into it he would find only pieces working upon one another, but never would he find anything to explain Perception."[2]

Leibniz invented a variety of machines, including clocks, lamps, pumps, propellers, submarines, and hydraulic presses. He built a mechanical calculator, the "stepped reckoner," which could add, subtract, multiply, and divide numbers with results up to sixteen digits. He believed that human reasoning could, in principle, be modeled by computational machines. But he saw no way for a machine to generate perceptual experiences.

The English biologist Thomas Huxley was flummoxed by this mystery in 1869: "How it is that anything so remarkable as a state of consciousness comes about as a result of irritating nervous tissue, is just as unaccountable as the appearance of the Djinn, when Aladdin rubbed his lamp."[3]

Huxley was an expert at anatomy and neuroanatomy. He compared the brains of humans and other primates, showing that the similarity of their structures supported Darwin's theory of human evolution. But he found nothing in the brain that could explain how it might generate conscious experiences.

The American psychologist William James grappled with the mystery of consciousness in 1890, exclaiming that "'A motion became a feeling!'—no phrase that our lips can frame is so devoid of apprehensible meaning." He agreed with the Irish physicist John Tyndall that, "The passage from the physics of the brain to the corresponding facts of consciousness is unthinkable."[4] Freud was confounded by the mystery: "We know two things concerning what we call our psyche or mental life: firstly, its bodily organ . . . and secondly, our acts of consciousness . . . so far as we are aware, there is no direct relation between them."[5] James and Freud offered deep insights into human psychology, and understood that psychology and neurobiology are correlated. But they had no theory of how brain activity might cause conscious experiences, no idea how to dispel the mystery.

Consciousness is still one of the great mysteries of science. A special 2005 issue of the journal *Science* ranked the top 125 open questions in science. The first-place winner was: *What is the universe made of?* A well-deserved win, given that today 96 percent of the matter and energy in the universe is "dark," meaning "we're in the dark about it."

The runner-up was: *What is the biological basis of consciousness?* This is the question that the Helmholtz Club pursued. It is the mystery that researchers around the world still struggle to solve.

Note how *Science* states the question: What is the *biological basis* of consciousness? It reveals the kind of answer that most researchers expect—that there is a biological basis for consciousness, that consciousness is somehow caused by, or arises from, or is identical to, certain kinds of biological processes. Given this assumption, the goal is to find the biological basis and describe how consciousness arises from it.

That there is a neural origin for consciousness was the working hypothesis of Francis Crick. As he put it, "The Astonishing Hypothesis is that 'You,' your joys and your sorrows, your memories and your ambitions, your sense of personal identity and free will, are in fact no more than the behavior of a vast assembly of nerve cells and their associated molecules. . . . 'You're nothing but a pack of neurons.' "[6]

This was the working hypothesis of the Helmholtz Club, and the reason that many of our invited speakers were, like Joe Bogen, experts in neuroscience. We sought clues that would lead us to the critical nerve cells and molecules that would crack the mystery of consciousness. Like paleontologists at a dig, we scoured the research of our speakers, hoping to unearth insights that could explain why some physical systems are conscious and others are not.

Our hope was not unfounded. For centuries, biologists sought a mechanism that would explain why some physical systems are alive and others are not. But vitalists, who hold that living organisms differ fundamentally from nonliving things, claimed that this quest would fail because, they argued, you cannot cook up life from the inanimate ingredients of the physical world; a special nonphysical ingredient, an élan vital, is also required. Debate between vitalists and biologists persisted until the celebrated discovery, in 1953, by James Watson and Francis Crick, of the double helix of DNA, which proved the vitalists wrong. This structure, with its four-letter code and penchant for replication, brilliantly solved the problem of cooking up life, mechanistically, from purely physical ingredients. It allowed the young field of molecular biology to wed naturally with Darwin's theory of evolution by natural selection—granting us tools to understand the evolution of life, to decipher its checkered odyssey over billions of years, and to create technologies that let us redesign life much as we please. The triumph of mechanistic physicalism over vitalism was decisive.

Inspired by this triumph, the Helmholtz Club expected that, in due course, consciousness would acquiesce to a mechanistic explanation couched in the language of neuroscience, opening new vistas for scientific exploration and technological innovation. In 1993, over lunch at the Club, Crick told me he was writing a book, *The Astonishing Hypothesis*, on neuroscience and consciousness. "Can you explain," I asked, "how neural activity causes conscious experiences, such as my experience of the color red?" "No," he said. "If you could make up any biological fact you want," I persisted, "can you think of one that would let you solve this problem?" "No," he replied, but added that we must pursue research in neuroscience until some discovery reveals the solution.

Crick was right. Absent a mathematical proof to the contrary, and given the impressive precedent of DNA, it is sensible to search for a double helix of neuroscience—a key fact whose discovery unravels the mystery of consciousness. It might be that our conscious web of dreams, aspirations, fears, sense of self, and sense of free will is spun by packs of neurons via a remarkable mechanism that we don't foresee. Our failure to envision a mechanism does not preclude one. Perhaps we're not clever enough, and an experiment will teach us what we can't surmise from an armchair. After all, we invest in experiments because they often repay us in surprise.

Consider, for instance, experiments on split-brain patients conducted by the neurobiologist Roger Sperry. They reveal several surprises about human consciousness. In one experiment, a person stares at a small cross in the center of a screen. Then two words, such as "KEY RING," flash on the screen for a tenth of a second, with "KEY" to the left of the cross and "RING" to the right—like this: KEY + RING

If you ask normal observers to report what they saw, they all say "key ring." The task is easy. A tenth of a second is plenty of time to read the words.

But if you ask split-brain patients, they say "ring." If you ask, "What kind of ring? A wedding ring, a doorbell ring, a key ring?" they stick with "ring." They cannot say what kind of ring.

You then blindfold a split-brain patient and bring out a box full of items: a ring, a key, a pencil, a spoon, a key ring, and so on. You ask the patient to reach in with their left hand and pick out the item that was named on the screen. Their left hand searches in the box, picking up and putting down items until it finds what it wants. When the left hand finally exits the box, it always holds a key. During its search, the left hand may encounter and reject a key ring.

After their left hand exits the box, you ask the blindfolded patient, "What's in your left hand?" They say they don't know. "Can you guess?" They guess small items that could fit in a box, such as a pencil or spoon. But they don't, except by accident, guess correctly.

You then ask the blindfolded patient to reach into the box with their right hand and retrieve the item that was named on the screen. Their right hand pulls out a ring. During its search, the right hand may encounter and reject a key ring. If you ask the blindfolded patient, "What's in your right hand?," they correctly and confidently say "ring."

Now, while the patient still holds an item in each hand, you remove the blindfold, let them see both hands, and ask, "You said you saw the word *ring*. So why does your left hand hold a key?" The patient either has no idea, or else confabulates, concocting a false story intended to be plausible. You then ask them, "Would you please draw with your left hand what you saw?" They draw a key.

Explaining experiments like these earned Roger Sperry a share of the Nobel Prize for Physiology and Medicine in 1981.

Sperry's explanation was simple and profound. When you fixate on the cross in KEY + RING, the neural pathways from eye to brain send KEY only to the right hemisphere, and RING only to the left. If the corpus callosum is intact, the right hemisphere then tells the left about KEY, and the left tells the right about RING, so that the person sees KEY RING.

If the callosum is cut, then the hemispheres no longer liaise. The right hemisphere sees KEY, the left sees RING, and neither sees KEY RING. The left can speak and the right cannot (apart from its talent to swear, which can become painfully apparent when a stroke in the left hemisphere leaves a person unable to speak but well able to turn the air blue). Thus, if the split-brain patient is asked, "What did you see?," the left hemisphere replies, "Ring."

The left hemisphere feels and controls the right hand. If the patient is asked, "Please pick out with your right hand what you saw," then the left hemisphere, guiding the right hand, picks what it saw: a ring.

The right hemisphere feels and controls the left hand. If the patient is asked, "Please pick out with your left hand what you saw," then the right hemisphere, guiding the left hand, picks what it saw: a key. When asked, "What's in your left hand?," the patient cannot say, because only the right hemisphere knows and only the left hemisphere speaks.

The "Astonishing Hypothesis" offers a cogent explanation: if consciousness arises from the interactions of a pack of neurons, then splitting that pack—and their interactions—can split consciousness.

To the untutored intuition, it seems unlikely that consciousness can be split with a scalpel. What could it mean to split my feelings, my knowledge, my emotions, my beliefs, my personality, my very self? Most of us would dismiss the idea as ludicrous. But to Sperry, after years of careful experiments, the evidence was clear: "Actually the evidence as we see it favors the view that the minor hemisphere is very conscious indeed, and further that both the separated left and the right hemispheres may be conscious simultaneously in different and even conflicting mental experiences that run along in parallel."[7]

The evidence for this conclusion has continued to mount. In one patient, the career goals of the two hemispheres differed: the left hemisphere said that it wanted to be a "draftsman," and the right hemisphere, using the left hand to assemble scrabble letters, wrote that it wished to "automobile race."[8] In another, the left hemisphere used the right hand to button a shirt, while the right hemisphere used the left hand to promptly unbutton it; the right hand lit a cigarette and the left put it out. Two persons, with distinct likes and dislikes, appear to reside—and sometimes quarrel—side by side, inside one skull.

Their differences can transcend the personal to the theological. In one patient studied by the neuroscientist V. S. Ramachandran, the pious left hemisphere believes in God, but the impious right does not.[9] When the bell tolls and both hemispheres approach the pearly gates, will Saint Peter need an assist from King Solomon? Or was the grim solution of Solomon already applied by the scalpel of Bogen? Tough questions for a future neurotheology.

What kind of creatures are we that our beliefs, desires, personalities, and perhaps the destinies of our souls can be split with a scalpel? Why are we conscious? What is consciousness? Can neuroscience decipher the perennial mystery of human consciousness? The searchlight of science, which has revealed insights into the realm of the impersonal—black holes, bound quarks, slow tectonic plates—is now being directed toward what matters to us most: our deeply personal world of conscious beliefs, desires, emotions, and sensory experiences. Might we glimpse and even comprehend our very selves? This is an aspiration of the science of consciousness.

Reaching this goal will require clever experiments and a soupçon of serendipity. Many experiments hunt for correlations between neural activity and consciousness, expecting that as the hunt succeeds, as the list of correlations grows, a critical discovery will solve the mystery of consciousness, just as the double helix solved the mystery of life.

We know that specific activities of the brain correlate with specific conscious (and unconscious) mental states. As we have discussed, activity of the entire left hemisphere, if surgically disconnected from the right, correlates with a repertoire of conscious states that is distinct from that of the right. But at finer levels of neural organization, we find a plethora of intriguing correlations.

For instance, activity in area V4 of the temporal lobe correlates with conscious experiences of color.[10] A stroke in V4 of the left hemisphere leads the patient to lose color in the right half of the visual world, a condition known as hemi-achromatopsia. If the patient stares, say, at the middle of a red apple, then the left half of the apple looks red and the right half looks gray. If, instead, a stroke damages area V4 in the right hemisphere, then the right half of the apple looks red and the left half looks gray.

A normal person can enter briefly into the color world of the hemi-achromatopsic via transcranial magnetic stimulation (TMS). TMS is induced by a strong magnet placed near the scalp, whose magnetic field is set either to enhance or impair activity in regions of the brain nearby. If TMS impairs activity of V4 in the left hemisphere, then, as the person watches, color drains from the right half of the world: if they look directly at a red apple, the right half of the apple fades to gray.[11] Turn off the TMS, and red color seeps back into the right half of the apple. If TMS stimulates V4, then the person will hallucinate "chromatophenes"—colored rings and halos.[12] With TMS, you can pour colors into consciousness, or siphon them out of consciousness.

Activity in a region of the brain called the postcentral gyrus correlates with conscious experiences of touch. The neurosurgeon Wilder Penfield reported in 1937 that stimulating this gyrus with an electrode in the left hemisphere prompted his patients to report conscious experiences of touch on the right side of the body; stimulating the right hemisphere led to feelings of touch on the left side of the body.[13] The correlation is systematic: nearby points on the gyrus correspond to nearby points on the body, and regions of the body that are more sensitive, such as the lips and fingertips, occupy more real estate on the gyrus. Stimulate the gyrus near the middle of the brain, and you feel it in your toes. Slide the electrode along the gyrus, stimulating at ever more lateral points, and the feeling, with a few exceptions, slides systematically up the body. The exceptions are interesting. The face, for instance, resides next to the hand on the gyrus. The toes are next to the genitals—a fact perhaps relevant to foot fetishes, as V. S. Ramachandran has suggested.[14]

Many experiments today continue the hunt for "neural correlates of consciousness" or NCCs.[15] This hunt is aided by a variety of technologies for measuring neural activity. For instance, functional magnetic resonance imaging (fMRI) tracks neural activity by measuring the flow of blood in the brain: neural activity, like muscle activity, requires a greater flow of blood to supply the extra energy and oxygen that are required. Electroencephalography (EEG), using electrodes glued to the scalp, tracks neural activity by measuring tiny fluctuations of voltage that it generates. Magnetoencephalography (MEG) tracks neural activity by measuring tiny fluctuations of magnetic fields. Microelectrodes can record the individual signals, called spikes or action potentials, of single neurons and small groups of neurons. Optogenetics uses colored lights to control and monitor the activity of neurons that have been genetically engineered to respond to specific colors.

The strategy of hunting for NCCs makes sense. If we want a theory that links neurons and consciousness, and we have no plausible ideas, then we can start by looking for correlations between them. Inspecting these correlations, we might discover a pattern that turns on a conceptual lightbulb. The path from correlation to causation, to be sure, is fraught with pitfalls: if a crowd forms at a train platform, then often a train soon arrives.[16] But crowds don't impel trains to roll in. Something else—a train schedule—creates the correlation between crowds and trains.

NCCs are key data for a theory of consciousness. Such a theory must perform two tasks. It must delineate the boundary between the conscious and unconscious, and it must explain the provenance and rich variety of our experiences—the taste of a lemon, the fear of spiders, the joy of discovery.

For the simpler (though not simple) task of demarcating the conscious and unconscious, we want to know how brain activity differs in the two cases. Here we have interesting data. For instance, in normal consciousness, neural activity is neither random nor too stable, but strikes a critical balance between the two—like a seasoned hiker that neither flits about nor loafs in one place, but intelligently explores the terrain. Propofol, which can induce general anesthesia, makes neural activity ploddingly stable.[17]

For the complex case of specific experiences—of tasting chocolate or fearing spiders—we want to find tight correlations between neural activity and each experience. But what is "tight"? That's not easy to nail down. Many researchers assume that it's the minimal neural activity that, under the right conditions, is sufficient to make the experience happen.[18] They search for this minimal activity by "contrastive analysis"—comparing how neural activity changes when an experience changes. For instance, if you view the "Necker" cube shown in Figure 1, you can have two different experiences. In one, face A is in front; in the other, face B. As you view the middle cube, you probably flip between the two experiences. A change in your neural activity that tracks your flip between experiences could be an NCC for your experience of the cube. The neat trick in this experiment is that your experience flips, but the image doesn't change. This makes it easier to ascribe your flip in conscious experience to the change in neural activity. But this activity still might not be the NCC. Some of the activity could be a precursor to the NCC, or a consequence of the NCC, rather than the NCC itself.[19] Careful experiments are required to tease these possibilities apart.

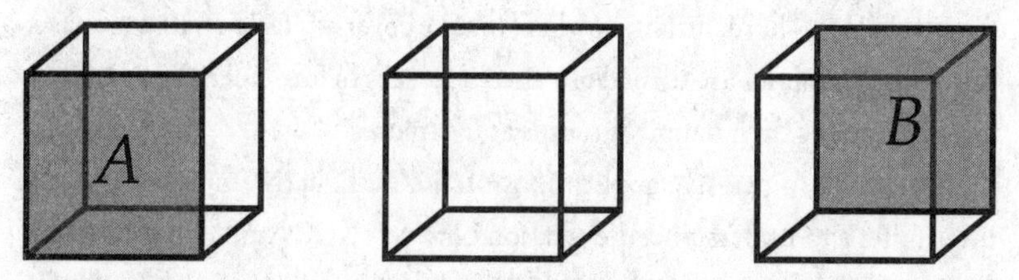

Fig. 1: The Necker cube. When we view the cube in the middle, we sometimes see face A in front, but at other times we see face B in front. © DONALD HOFFMAN

NCCs are important for theory, and also for practice. Arachnophobia, an excessive fear of spiders, is correlated with activity in the amygdala. Triggering this fear and its NCC in the amygdala allows both to be erased. Merel Kindt, a psychotherapist in the Netherlands, cures arachnophobia by asking the arachnophobe first to touch a live tarantula, thus activating the phobia and its NCC. She then asks the patient to take a forty-milligram pill of propranolol, a β-adrenergic blocker that disrupts the NCC from being stored back into memory. When the patient returns the next day, the phobia is gone.[20] This therapy holds promise for other phobias, and for posttraumatic stress disorder.

Another example exploits optogenetics, a biological technique that uses light to control neurons that have been genetically altered. With optogenetics, it's now possible to turn on an NCC for a positive feeling at the flip of a switch and then, just as quickly, to turn it off. Christine Denny, at Columbia University, has pulled off this remarkable feat using mice genetically engineered with a gene from algae that codes for a light-sensitive protein.[21] In nature, the algae use this protein to respond intelligently to light. In the engineered mouse, the gene hides silently, unexpressed, until the drug tamoxifen is injected. Then, for a brief time, any neurons that happen to become electrically excited will activate the gene and insert the protein into their membranes. Denny places an injected mouse into an environment it likes: soft, dim, with places to take cover. The mouse happily explores this idyllic environment, and any neurons engaged in creating a happy NCC insert the protein into their membranes. Then later, Denny can trigger its happy NCC using fiber optics that flash into its brain a colored light that activates the protein. Even if the mouse sits in a frightful place—hard, bright, nowhere to take cover—it feels a halcyon space, until the fiber optics are turned off. Then it freezes in fear. Turn the light back on, and once again it happily grooms and explores.

These are impressive applications of NCCs. Equally impressive is our utter failure to understand the relation between NCCs and consciousness. We have no scientific theories that explain how brain activity—or computer activity, or any other kind of physical activity—could cause, or be, or somehow give rise to, conscious experience. We don't have even one idea that's remotely plausible. If we consider not just brain activity, but also the complex interactions among brains, bodies, and the environment, we still strike out. We're stuck. Our utter failure leads some to call this the "hard problem" of consciousness, or simply a "mystery."[22] We know far more neuroscience than Huxley did in 1869. Yet each scientific theory that tries to conjure consciousness from the complexity of interactions among brain, body, and environment always invokes a miracle—at precisely that critical point where experience blossoms from complexity. The theories are Rube Goldberg devices that lack a critical domino and need a sneak push to complete the trick.

What do we want in a scientific theory of consciousness? Consider the case of tasting basil versus hearing a siren. For a theory that proposes that brain activity *causes* conscious experiences, we want mathematical laws or principles that state precisely which brain activities cause the conscious experience of tasting basil, precisely why this activity does not cause the experience of, say, hearing a siren, and precisely how this activity must change to transform the experience from tasting basil to, say, tasting rosemary. These laws or principles must apply across species, or else explain precisely why different species require different laws. No such laws, indeed no plausible ideas, have ever been proposed.

If we propose that brain activity is identical to, or gives rise to, conscious experiences, then we want the same kind of precise laws or principles—that link each specific conscious experience, such as the taste of basil, with the specific brain activities that it is identical to, or with the specific brain activities that give rise to it. No such laws or principles have been offered.[23] If we propose that conscious experience is identical, say, to certain processes of the brain that monitor other processes, then we need to write down laws or principles that precisely specify these processes and the conscious experiences with which they are identical. If we propose that conscious experience is an illusion arising from some brain processes attending to, monitoring, and describing other brain processes, then we must state laws or principles that precisely specify these processes and the illusions they generate. And if we propose that conscious experiences emerge from brain processes, then we must give the laws or principles that describe precisely when, and how, each specific experience emerges. Until then, these ideas aren't even wrong. Hand waves about identity, emergence, or attentional processes that describe other brain processes are no substitute for precise laws or principles that make quantitative predictions.

We have scientific laws that predict black holes, the dynamics of quarks, and the evolution of the universe. Yet we have no clue how to formulate laws, principles, or mechanisms that predict our quotidian experiences of tasting herbs and hearing street noise.

Perhaps Crick was right: maybe we just haven't found the crucial experiment that unveils the breakthrough idea. Perhaps one day—funding permitting—we will: the double helix of neuroscience will be discovered, and a genuine theory of consciousness will follow.

Or perhaps we were short-changed by evolution, and lack the concepts needed to understand the relationship between brains and consciousness. Cats can't do calculus and monkeys can't do quantum theory, so why assume that *Homo sapiens* can demystify consciousness? Perhaps we don't need more data. Perhaps what we need is a mutation that lets us understand the data we have.

Noam Chomsky dismisses arguments from evolution about limits to our cognitive capacities. But he insists nonetheless that we must recognize "the scope and limits of human understanding" and that "some differently structured intelligence might regard human mysteries as simple problems and wonder that we cannot find the answers, much as we can observe the inability of rats to run prime number mazes because of the very design of their cognitive nature."[24]

I suspect Chomsky is right: there are limits to human understanding. And I admit that these limits, whether they derive from evolution or another source, may preclude us from understanding the relation between consciousness and neural activity.

But before punting the hard problem of consciousness, we might consider a different possibility: perhaps we possess the necessary intelligence and are hindered by a false belief.

False beliefs, rather than innate limits, can stump our efforts to solve puzzles. Examples of this are standard fare in textbooks on cognitive science. In one example, people are given a candle, a box of thumbtacks, and a book of matches. They're asked to fasten the candle to a wall so that, when lit, its wax can't drip on the floor. Most people fail. They tacitly assume that the box must do one thing—hold thumbtacks. They don't think to dump the tacks out of the box, to use the tacks to fasten the box to the wall, and to put the candle in the box. To solve the puzzle, they must challenge a false assumption.

What false assumption bedevils our efforts to unravel the relation between brain and consciousness? I propose it is this: *we see reality as it is.*

Of course, no one believes that we see *all* of reality as it is. Physicists tell us, for instance, that the light we see is a tiny fraction of an immense electromagnetic spectrum that we can't see—including ultraviolet, infrared, radio waves, microwaves, X-rays, and cosmic rays. Some animals perceive what we cannot: birds and bees see ultraviolet; pit vipers "see" infrared; elephants hear infrasound; bears smell distant carcasses; sharks "feel" electric fields; pigeons navigate by magnetic fields.

But most of us believe that, in the normal case, we accurately see *some* of reality as it is. Suppose I open my eyes and have a visual experience that I describe as a red tomato a meter away. Then I close my eyes and my experience changes to a mottled gray field. If I'm sober and healthy, and don't think I'm being tricked, then I believe that even when my eyes are closed, even while I experience a gray field, nevertheless there really is a red tomato a meter away. When I open my eyes and again have an experience that I describe as a red tomato a meter away, I take this as evidence that the tomato was there all along. To gather further evidence for my belief, while my eyes are closed I can reach out and feel the tomato, lean over and smell it, or ask a friend to look and confirm that it's still there. The convergence of all this evidence convinces me that a real tomato is indeed there even when all eyes are closed and no hand touches it.

But could I be wrong?

This question, I admit, sounds faintly mad. Most sane persons, given this evidence, would surely conclude that the tomato is still there. Its existence when unseen and untouched seems to be an obvious fact, not a misguided belief.

But this conclusion is a fallible belief, not a dictate of logic or an indubitable fact. We must test its validity against advances in fields such as cognitive neuroscience, evolutionary game theory, and physics. When we do so, the belief proves false.

This surprising result is the subject of this book. I don't try to solve the mystery of consciousness. But I do try, in the coming chapters, to dethrone a belief that hinders a solution. In the last chapter, I suggest how we may tackle the mystery of consciousness once we have shed the burden of this false belief.

What could it mean to claim that no tomato is there when I don't look? Our intuitions here can be helped by a glance back at the Necker cube. As we discussed, you can see a cube with face A in front—call it Cube A. Or you can see a cube with face B in front—call it Cube B. Each time you view the figure you see Cube A or Cube B, but never both at once.

When you look away, which cube is there: Cube A or Cube B?

Suppose you saw Cube A just before you looked away, and you answer that Cube A is still there. You can check your answer by looking back. If you do this a few times, you'll discover that sometimes you see Cube B. When this happens, did Cube A transform into Cube B when you looked away?

Or you can check your answer by asking friends to look. You'll find that they often disagree, some saying that they see Cube A, others that they see Cube B. They may all be telling the truth, as you could check with a polygraph.

This suggests that neither Cube A nor Cube B is there when no one looks, and that there is no objective cube that exists unobserved, no publicly available cube waiting for all to see. Instead, if you see Cube A while your friend sees Cube B, then in that moment you each see the cube that your visual system constructs. There are as many cubes as there are observers constructing cubes. And when you look away, your cube ceases to be.

This example is meant only to illustrate what it may mean to say that no tomato is there when you look away. It does not, of course, prove that no tomato is there when you look away. After all, one could argue, the Necker cube is illusory but a tomato is not. Making the case against unseen tomatoes is not trivial. The core point will be that the reality prompting you to create your experience of a tomato is nothing like what you see and taste. We have been misled by our perceptions.

In fact, we have a long history of being misled. Many ancient cultures, including the pre-Socratic Greeks, were misled by their perceptions to believe that the earth is flat. It took the genius of Pythagoras, Parmenides, and Aristotle to discover, despite the testimony of the eye, that the earth is roughly a sphere. For many centuries after this discovery, most geniuses, with the exception of Aristarchus (ca. 310 BC–ca. 230 BC), were misled by their perceptions to believe that our spherical earth is the unmoving center of the universe. After all, apart from earthquakes, the earth never appears to move; and it looks as if the sun, stars, and planets circle the earth. Ptolemy (ca. 85–ca. 165) built this geocentric misreading of perception into a model of the universe that, according to the Catholic Church for fourteen centuries, brandished the imprimatur of Holy Scripture.

Our penchant to misread our perceptions, as philosopher Ludwig Wittgenstein pointed out to his fellow philosopher Elizabeth Anscombe, stems in part from an uncritical attitude toward our perceptions, toward what we mean by "it looks as if." Anscombe says of Wittgenstein that, "He once greeted me with the question: 'Why do people say that it was natural to think that the sun went round the earth rather than that the earth turned on its axis?' I replied: 'I suppose, because it looked as if the sun went round the earth.' 'Well,' he asked, 'what would it have looked like if it had *looked* as if the earth turned on its axis?' The question brought it out that I had hitherto given no relevant meaning to 'it looks as if' in 'it looks as if the sun goes around the earth.' "[25] Wittgenstein's point is germane any time we wish to claim that reality matches or mismatches our perceptions. There is, as we shall see, a way to give precise meaning to this claim using the tools of evolutionary game theory: we can prove that if our perceptions were shaped by natural selection then they almost surely evolved to hide reality. They just report fitness.

In 1543, Copernicus's book *De revolutionibus orbium coelestium* (*On the Revolutions of the Celestial Spheres*) was published posthumously. In it, he proposed, as Aristarchus had before, that the earth and other planets go around the sun. Galileo peered through a telescope and saw evidence for this theory—moons orbiting Jupiter, and Venus changing phases, like our moon. The Church opposed this theory and tried Galileo in 1633 for heresy, for his temerity to claim "that one may hold and defend as probable an opinion after it has been declared and defined contrary to the Holy Scripture." Galileo was forced to recant, and sentenced to house arrest for the remainder of his life. It wasn't until 1992 that the Church acknowledged its error.

Several factors contributed to this error. One was belief in the idea of a Great Chain of Being—with God and the perfection of celestial spheres above, and man and the imperfection of the sublunary realm below—that comported well with the Ptolemaic system.[26] But a key factor was a simple misreading of our perceptions: the Church thought we can just see that the earth never moves and is the center of the universe.

As noted in the epigraph to this book, Galileo argued that we misread our perceptions in other ways: "I think that tastes, odors, colors, and so on are no more than mere names so far as the object in which we locate them are concerned, and that they reside in consciousness. Hence if the living creature were removed, all these qualities would be wiped away and annihilated."[27] We naturally think that a tomato is still there—including its taste, odor, and color—even when we don't look. Galileo disagreed. He held that the tomato is there, but not its taste, odor, and color—these are properties of perception, not of reality as it is apart from perception. If consciousness disappeared, so would they.

But he thought the tomato itself would still exist, including its body, shape, and position. For these properties, he claimed, we see reality as it is. Most of us would agree.

But evolution disagrees. We will see in chapter four that evolution by natural selection entails a counterintuitive theorem: *the probability is zero that we see reality as it is.* This theorem applies not just to taste, odor, and color, but also to shape, position, mass, and velocity—even to space and time. We see none of reality as it is. The reality that prompts you to create an experience of a tomato, the reality that exists whether or not you see a tomato, is nothing like what you see and taste.

We discarded a flat earth and a geocentric universe. We realized that we had misread our perceptions, and we corrected our errors. It wasn't easy. In the process, mundane intuitions and Church doctrines were shattered. But these corrections were mere warm-ups. Now we must jettison spacetime itself, and everything in it.

What kind of creatures are we? According to evolution, not creatures that see reality as it is. And that profoundly affects how we think about the relation between brains and consciousness. If space and time exist only in our perceptions, then how can anything within space and time, such as neurons and their activity, create our consciousness?

Understanding the evolution of perception is a critical step toward understanding who we are, and the provenance of our consciousness.

CHAPTER TWO

Beauty

Sirens of the Gene

"In the distant future I see open fields for far more important researches. Psychology will be based on a new foundation."

—CHARLES DARWIN, *ON THE ORIGIN OF SPECIES*

"Good Lord Boyet, my beauty, though but mean,
Needs not the painted flourish of your praise:
Beauty is bought by judgement of the eye,
Not utter'd by base sale of chapmen's tongues"

—SHAKESPEARE, *LOVE'S LABOUR'S LOST*

In 1757, David Hume argued in his book *Standard of Taste* that beauty is in the eye of the beholder. "Beauty," he said, "is no quality in things themselves: It exists merely in the mind which contemplates them; and each mind perceives a different beauty." This naturally raises a question: Why is *this* standard of beauty in the eye of *that* beholder? A century after Hume, Darwin laid the foundation—evolution by natural selection—for a psychology that explains why: beauty is a perception of fitness payoffs on offer, such as the payoff for eating that apple or dating that person. This perception will differ—from species to species, person to person, and even time to time—as needs and niches differ. Reproductive success depends on collecting fitness points. Beauty tells us what and where they are.

Evolutionary psychology makes new, and surprising, predictions about our judgments of human beauty. Each time, for instance, that you glance at a face, you scrutinize its eyes—scoring them on a checklist of details—and arrive, through unconscious deliberation, at a verdict on their beauty. What women find attractive about the eyes of a man sometimes differs from what men find attractive about the eyes of a woman. Our ancestors relied on this unwritten checklist for millennia, but the new science of beauty has revealed some of its items. We discuss these items and the logic of their discovery, as well as some practical applications.

The predictions of evolution about beauty are surprising but, as we will see in chapter nine, its predictions about physical objects are disconcerting: objects, like beauty, are in the eye of the beholder and inform us about fitness—not about objective reality. To prepare us for the perplexing case of objects, let's warm up our intuitions by exploring the perception of beauty in the animal kingdom.

Male jewel beetles, *Julodimorpha bakewelli,* have a thing for beautiful females.[1] The males fly about, searching for females, which are shiny, dimpled, and brown. Recently, some male primates of the *Homo sapiens* species have been driving through the beetle's haunts in Western Australia and littering the outback with emptied beer bottles, known as "stubbies." As it happened, some of the stubbies were shiny, dimpled, and just the right shade of brown to catch the fancy of male beetles. Forsaking real females, the male beetles swooned over stubbies with their genitalia everted, and doggedly tried to mate despite glassy rebuffs. (A classic case of the male leaving the female for the bottle.) Adding injury to insult, ants of the species *Iridomyrmex discors* learned to loiter near stubbies, wait for the befuddled and priapistic beetles, and then devour them, genitalia first, as they failed to have their way.

The poor beetles teetered on extinction, and Australia had to change its beer bottles to save its beetles.

This blunder of the beetle is surprising. Male beetles have mated with females for untold thousands of years. You would think that they surely know their females. Apparently not. Even when a male crawls all over his stubby—enjoying full embodied contact—he perceives it as a Siren, a 370-milliliter Amazon of irresistible allure.

Something is awry. Why should a beetle fall for a bottle? Is it due, perhaps, to his tiny brain? Perhaps mammals, with their bigger brains, would never make such a silly mistake. But they do. Moose in Alaska, Montana, and elsewhere have been found, and photographed, mating with metal statues of moose, and even bison, sometimes for hours on end. We can laugh, but *Homo sapiens* has its own checkered history, including sex dolls that starred centuries ago in Mughal paintings of India, and robots that star today in the International Congress on Love and Sex With Robots. Our bigger brains guarantee no inerrant attraction to bona fide human beauties.

What, then, is beauty? Surprisingly, given the panoply of foibles besetting beetles, moose, *Homo sapiens*, and many other species, beauty is the intelligent verdict of a complex but mostly unconscious computation. Each time you encounter a person, your senses automatically inspect dozens, perhaps hundreds, of telltale clues—all in a fraction of a second. These clues, meticulously selected through eons of evolution, inform you about one thing: reproductive potential. That is, could this person have, and raise, healthy offspring? Of course, explicit thoughts about this question, and explicit clues to a verdict, are not what you typically experience in that encounter. Instead you experience just the verdict itself—as a feeling that varies from hot to not. That feeling, that executive summary of a painstaking investigation, is the beauty in the eye of the beholder.

Which gives the lie to the idea that beauty is a *whim* of the beholder. To the contrary, it is the consequence of unconscious inferences within the beholder, inferences that were crafted over millennia by the logic of natural selection: if the inferences too often delivered a verdict of hot when they should have said not, or vice versa, then the beholder would too often prefer mates who were less likely to raise healthy offspring. In this case, the beholder's misguiding genes, and their faulty inferences, would be less likely to pass into the next generation. In short, if genes get beauty wrong, they tend to go extinct. This is the pitiless logic of natural selection.

It's all about struggles between genes. Which is to say, it's all about *fitness*—the central concept of evolution by natural selection. Genes that are more adept at elbowing their way into the next generation are said to be *fitter.* Even a slight excess of talent in the art of the elbow can allow a gene to proliferate across generations and eradicate competitors of but moderate talent. Oscar Wilde understood this logic well. "Moderation," he wrote, "is a fatal thing. . . . Nothing succeeds like excess."[2]

Genes don't elbow each other directly. They do it by proxy. They boot up bodies and minds—phenotypes—and let them duke it out. Phenotypes that fare better at the brawl are, like their respective genotypes, said to be fitter. The fitness of a phenotype depends, of course, not just on genes, but also on the vagaries of disease, development, nutrition, and the common depredations of time. Identical twins, for instance, can differ in their phenotypic fitness. But make no mistake: even though genes battle by proxy, they have skin in the game. Like pilots in a plane, genes sit strapped into their phenotype: if it crashes, they die.

The computation of beauty is part of the battle by proxy, one of the ingenious devices deployed by genes to compete with other genes—to enhance fitness. Your computation of beauty, in a recursive twist, can enhance your own fitness if you compute beauty better than your competition does. Fitness—enhancing it, estimating it, and enhancing it by estimating it—is the preoccupation of evolution by natural selection. The computation of beauty is wired into us early in life. Infants as young as two months of age look longer at faces that adults rate more attractive.[3]

The trouble with computing beauty, with ferreting out the fitness of genes, is that genes themselves are invisible. This forces genes to hunt for evidence of fitness in the only place where it can be seen—in phenotypes, in the bodies and minds that other genes have fashioned and pressed into their service. But a phenotype rarely wears its fitness on its sleeve; it must be scoured for clues.

Sherlock Holmes claimed that the success of a detective depends on "the observation of trifles."[4] One trifle in the search for beauty is a feature of the human eye called the *limbal ring*, a dark annulus at the border between the colored iris and the white sclera. I first noticed this ring in the *Afghan Girl*, a photograph of Sharbat Gula that graced the June 1985 cover of *National Geographic* and became the most recognized photograph in the magazine's history.[5] I wondered whether her prominent limbal rings, which transform her eyes into veritable bull's-eyes, might rivet our attention and enhance her beauty.

Why might prominent limbal rings be attractive? Or, to ask this in the language of evolution: Why might such rings signal greater fitness?

As it happens, prominent rings signal health. For limbal rings to be prominent they must be visible, and, for that, the cornea—the transparent outer layer of the eye—must be clear and healthy. Diseases such as glaucoma and corneal edema can cloud the cornea, making limbal rings less visible. Poor lipid metabolism can trigger arcus senilis, milky deposits of cholesterol that hide the rings. Disregulation of calcium in the blood can cause limbus sign, milky deposits of calcium that, again, hide the rings. A medley of diseases can obscure the limbal rings; someone with distinct rings is less likely to suffer them.

Prominent rings also signal fitness by signaling youth. Measurements by Darren Peshek, then a graduate student in my lab, assisted by a team of undergraduates, found that the thickness of limbal rings, and hence their prominence, declines with age.[6]

In principle, then, limbal rings signal youth, health, and thus fitness. But has evolution in fact tuned our hot-or-not meter, the computation of beauty within the *Homo sapiens* beholder, to spot the subtle clues to fitness in limbal rings?

To find out, Peshek showed observers on each trial of an experiment a pair of faces that were identical, except that one had limbal rings and one did not. Observers had to pick the face that looked more attractive. The data were clear: male and female observers prefer male and female faces with limbal rings, even if the faces are shown upside down.[7] Then, through a sequence of experiments, Peshek discovered the ideal rings—those whose thickness, opacity, and tapering look most attractive.[8]

Knowing this ideal, you can enhance your portrait by editing your rings, or kick up your eyes with contacts, now available, that mimic hot rings—like makeup applied directly to the eye itself.

This highlights a hazard for beholders of beauty: genes can lie about fitness. They can rig their phenotype—planting mendacious clues in its body and deceptions on its mind. By lying about the fitness that they offer a beholder, genes can amass more fitness for themselves.

Sometimes the lie is white. Lipstick and eyeliner have never hurt a soul.

Sometimes the lie is cynical and exploitative. Hammer orchids, of the genus *Drakaea* in Western Australia, peddle sex to thynnid wasps.[9] The female wasp, when in the mood, climbs a blade of grass and rubs her legs to broadcast a scent appealing to males. A charmed male tracks her scent and flies a snaking pattern upwind until he finds her. He embraces her, whisks her up to the meter-high club, then down to his prearranged pad, which caters a gourmet banquet of beetle larvae. There she lays her eggs and dies.

The average flower next door has no chance to seduce a male thynnid. But the genes of the hammer orchid have given it a celebrity makeover: a green and slender stem with the ambience of grass; dangling from its top, a labellum with the shapely curves, alluring color, velvety texture, and enticing scent of a female thynnid. An entranced male tries to whisk off with the labellum, but learns that this would-be mate will not cooperate. He eventually flies off in frustration, bearing pollen daubed on him surreptitiously during his deflating ordeal. When he tries his luck with another fake mate, he pollinates it. In this charade, *Drakaea* genes get fitness; the wasp just gets used.

The lies of genes in the quest for fitness can cross the border from cynical to sinister. Female fireflies of the genus *Photuris* lure male fireflies of the genus *Photinus*—with a tragic ending.[10] On a lonely night, a *Photinus* male emits a sequence of flashes. A receptive *Photinus* female can answer with a sequence of flashes that dovetail with his to form a choreographed duet. Upon receiving her reply, the hopeful male flies to her and mates.

The *Photuris* female has broken the code of *Photinus* and responds to a *Photinus* male's flashes with the proper duet. When the *Photinus* male arrives for his tryst, he finds a female much larger than he expected, and gets eaten.

The callous genes of *Photuris* promise *Photinus* the ultimate in fitness rewards, but deliver instead the ultimate in fitness penalties. This sinister bait-and-switch enhances the fitness of *Photuris* in an obvious way—vital calories—but with a less conspicuous twist: *Photinus* fireflies contain lucibufagins (LBGs), steroids toxic to many potential predators. When bitten or squeezed, a *Photinus* firefly exudes a drop of blood laden with LBGs that, to a would-be predator, tastes foul (meaning "bad for my fitness"), prompting it to release the firefly. The *Photuris* firefly, by eating a *Photinus* laden with LBGs, innoculates itself against predators.

Beauty is our best estimate of reproductive potential. But as the sagas of *Photuris* and *Drakaea* and countless others reveal, the genes behind the scenes of the beauty game are ruthless operators, unfettered by moral compunction, unhesitant to deceive and destroy in their single-minded quest to enhance their own fitness—to amass fitness points. They play for keeps in zero-sum games. *Photuris* devours *Photinus* and racks up fitness points by siphoning all of its calories and LBGs; *Photinus* loses everything. *Drakaea* deceives a thynnid and racks up fitness points in the form of pollination; the thynnid loses fitness points in the form of time and calories wasted on *Drakaea*. Fitness points are the coin of the realm: the more one collects, the greater one's chance to succeed in reproduction. Machiavellian genes nab fitness points, not as honest wages, but as filthy lucre.

Fitness points are not carved in stone, but are as varied as the organisms that seek them and as fickle as the desires that signal them. For a *Photinus* male looking to mate, an eligible *Photinus* female offers a fitness bonanza; for an amorous male of *Homo sapiens,* she offers nothing. A change of organism, with all else fixed, can radically change the fitness payoffs.

The payoffs to an organism vary with its state. A clear example is hunger.

The delight of a famished teen at the smell of a pizza signals the bounty of fitness offered by the first slice. The indifference, or even disgust, of that teen an hour and six slices later to that same smell signals a dearth of fitness. Same teen, same pizza, but a big change in the fitness on offer because the state and needs of the teen have changed. Fitness points depend on the organism, its state, and its action.

Your feeling of sexual attraction, from hot to not, signals your sophisticated estimate of reproductive potential. This estimate, we have seen, heeds the state of the limbal ring. What other features of the eye, I wondered, might it attend? Flipping through photos of faces, I noticed that the colored iris looked larger in the eyes of infants than of adults. Negar Sammaknejad, a former graduate student in my lab assisted by undergraduates, confirmed and refined my informal observation with careful measurements on a database of photographs: from birth to age fifty, there is a decline in the area of the iris relative to the white sclera; but from age fifty on, this area of the iris increases, as tissues around the eyes sag and cover the sclera.[11] So the area of the iris, relative to the sclera, varies systematically with age.

These data led me to predict that men prefer, in women under fifty, irises that are slightly larger. The facts underwriting this prediction are simple: larger irises, and fertility, correlate with youth in females under fifty. The infertility rate for females aged twenty is about 3 percent; aged thirty, about 8 percent; aged forty, about 32 percent; aged fifty, 100 percent. The likelihood of success in getting pregnant for females aged twenty is about 86 percent; at age thirty, it is about 63 percent; at age forty, it is about 36 percent; and at age fifty, it is about zero.[12]

This decline in female fertility has shaped, through natural selection, male judgments of female beauty. The logic is simple: consider a man whose genes happen to code for a computation of beauty that prizes women over, say, age fifty. He can enjoy life in the company of these beauties. But what is the chance that they will bear children with his genes and his computation of beauty? Almost none. By contrast, what is this chance for a man whose genes prize women age twenty? Nearly certain.

There is, however, a twist: a woman's fertility is not the same as her reproductive value—the number of offspring she can expect in the future. Genes that prize reproductive value tend to win, to elbow their way into the next generation. This value peaks at age twenty. A woman at twenty-five may be more fertile than she was at twenty, but her reproductive value was greater at twenty.[13]

So we expect that natural selection has shaped men to find women most beautiful at about twenty. This leads to a clean prediction: men over twenty should prefer younger women; men under twenty should prefer *older* women.

Both predictions have been confirmed in experiments. Men over twenty prefer younger women. No surprise. But teen males prefer women who are slightly *older*.[14] This supports an evolutionary account over certain rival accounts. The preference of teen males is not, for instance, due to positive reinforcement from older women, who rarely reciprocate teen advances. It is not a desire to dominate, which is unlikely to succeed with older women. Nor is it due to culture; the experiments have been replicated in several cultures.

In sum, natural selection fashioned within men a feeling for beauty that pivots on evidence of reproductive value. Any signal of youth, such as a larger iris, is crucial evidence of reproductive value in a woman. So I predicted, in 2010, that men prefer larger irises in women under fifty. This prediction is distinct from the prediction about the attractiveness of limbal rings; the size of an iris can vary without varying the size or visibility of its limbal ring.

To test this prediction, Sammaknejad showed observers pairs of faces that were identical, except that the irises of one face were larger.[15] Observers picked the more attractive face. The data were clear: men prefer female faces with larger irises, even if the faces are seen upside down.[16]

Our genes compel men to detect and desire this subtle cue of female fitness. A woman who knows this can enhance her beauty: in photographs, she can simply edit her irises; in daily life, she can wear "big eyes" contacts that enlarge irises. These contacts are now popular in Japan, Singapore, and South Korea. An artist who understands the impact of iris size can manipulate her viewers. Indeed art, in this case, anticipates science: Japanese anime and manga cartoons, seeking to accentuate youth, depicted female characters with large irises long before our research.

What about women? Do they prefer large irises in men? Recall that a limbal ring signals youth and health by being distinct, and that women evolved to prefer men with distinct rings. But a large iris only signals youth; unlike a distinct ring, which bespeaks an eye that is clear and thus free of disease, a large iris offers little clue to health, other than the clue of youth. So, in the case of irises, unlike the case of limbal rings, it is more difficult to predict what women want. Their tastes are more complex.

This complexity of preference is for a good evolutionary reason: parental investment. Raising offspring demands some investment of time and energy from each parent, but the amount of investment can differ between the two parents. In mammals, the female must invest heavily, in gestation and nursing. The male, however, may invest heavily, providing food and protection, or minimally, by simply mating and leaving.

The greater your investment, the fussier your choice of mate.[17] If each mating is costly, then you will choose judiciously: genes that code for rash choices are less likely to survive into the next generation. If, however, your investment is small, then another strategy is available: be less picky and have multiple mates. Genes that adopt this strategy of quantity over quality can still perpetuate themselves across generations, even if each offspring has less chance to survive.

The sex with greater investment is pickier in choosing mates. The one with less investment is less choosy and competes for access to the pickier sex—in some cases with physical battles, and in other cases, such as the peacock, with impressive displays. This explains why, typically, men court and women choose.

However, the investments, and thus these roles, are reversed in some species. For certain sea horses, the males are the keepers of the bag of eggs; in this case the females court and the males choose.[18]

In species where the sexes have equal investment, both are finicky. The crested auklet, for instance, is a seabird dwelling in the northern Pacific and Bering Sea.[19] A mating pair has a single offspring, which both parents equally incubate as an egg and raise as a chick. Both sexes sport colorful plumage and a forehead crest, exude a strong citrus scent, and boast a complex trumpet call.

Human biology dictates that each woman must invest heavily in each child. But it gives each man a choice. Some men invest little. But many choose to invest heavily, to provide food and protection for their mate and children. In no other species of primates do males regularly provide food; females fend for themselves.[20]

A woman who mates with a man of resources and commitment will more likely succeed in raising kids. So selection shaped women to prefer men with resources and with status, which correlates with resources. This preference crosses cultures and intensifies in women who have more resources. It is no side effect of financial inequality.[21] A man's age and height correlate with his status and resources; women, across cultures, prefer tall and slightly older men.[22] A woman can tell, from a photo of a face, if a man is prone to cheat and divert resources to other women; cheaters tend to look more masculine, but not more attractive.[23] Men are less able to discern female cheaters.[24] Indeed, as moose and beetles demonstrate, males with little investment sometimes fail to discern females from bottles or statues.

A woman who mates with a man of good genes will more likely succeed in raising healthy kids. Such genes correlate with levels of testosterone.[25] Because testosterone promotes the growth of bone and muscle, men with more testosterone during puberty develop more masculine faces with longer and squarer jaws, and larger eyebrow ridges. So selection shaped women to prefer men with more masculine faces. But there is a wrinkle: higher testosterone is correlated with less investment in offspring and a greater tendency to cheat.[26]

A woman faces a fitness tradeoff: mate with a man of lower testosterone but higher commitment, or mate with a man of higher testosterone but lower commitment. Tradeoffs like this are common in evolution, and genes that play the tradeoff better will more often get the nod to the next generation. In the case of women, the genes are geniuses, and strive to reap the fitness benefits of both choices: they incline women to prefer masculine faces more strongly in the high fertility phase of the menstrual cycle.[27] They choreograph hormones and brain activity to shift a woman's desires for male faces throughout the monthly cycle,[28] increasing the chance that her kids will have good genes and a committed man.

But genes don't stop at masculine faces. They choreograph a woman's preference for masculine gaits, bodies, odors, voices, and personalities.[29] Women in the low fertility phase feel more commitment to their partner, but during the high fertility phase they are more prone to cheat, to fantasize about cheating, to dress attractively, and to meet and flirt with new men.[30] If, however, a woman's partner is attractive, or if his MHC genes, which code for the immune system, complement hers and incline their children toward immune health, then her wandering eye is less pronounced—again a clever strategy by genes to play the odds for a greater fitness payoff.[31] For the most part, these machinations of genes fly under the radar of conscious experience and foster, but do not force, a choice of action.

Given these unconscious intrigues of unscrupulous genes, it is tricky to predict what a woman might want in the iris of a man. A smaller iris suggests greater age and thus greater resources. A larger iris suggests youth and thus healthier genes. Perhaps a woman prefers a smaller iris when her fertility is low, and a larger iris when it is high. Sammaknejad's experiment did not measure fertility, and found no preference for iris size, perhaps because her data averaged differing preferences over the course of a cycle.

At the center of the iris is a pupil, an opening that lets light pass into the eye. The pupil dilates and constricts as the ambient light dims and brightens. But the pupil also dilates in response to cognitive states, such as interest or mental effort, and to emotional states, such as fear or attraction.[32] As we age, the maximum dilation of the pupil declines.[33]

When a man sees a woman with a smile and large pupils, he also unconsciously sees interest. As you may expect from the sex with lower parental investment, he finds this attractive.[34] In one experiment, a book was sold whose cover bore the face of a smiling woman. On some covers, her pupils were artificially enlarged. Men preferred to buy a book with larger pupils, though they could not say why.[35] They picked up a genuine, albeit fallible, clue of a woman's interest: the pupils of a woman will, when her fertility is high, dilate more to a sexually arousing image—unless she is on the birth control pill.[36]

In her first experiment, Sammaknejad darkened irises so that pupils were not visible and influential. But in a second experiment, she studied how the sizes of iris and pupil interact to influence attraction.[37] She showed men, on each trial, two photos of a woman's face that were identical, except that one had larger irises and pupils. The men were asked to pick the more attractive face. As expected, they picked the face with larger irises and pupils: these are cues to youth and interest. Then Sammaknejad put the men in a quandary. On each trial, she showed them two photos of a woman's face that were identical, except that one had larger irises and smaller pupils. This forced a man to choose between a "younger" woman showing less interest and an "older" woman showing more. Different men took different strategies: some chose the younger face, others the face showing interest. Such variations of strategy are green shoots for the pruning hand of natural selection.

When in low fertility, women prefer smaller pupils—less interest—in the eyes of men. A few days before ovulation, they switch to prefer larger pupils.[38] This early switch might have evolved to allow them time to create and evaluate a short list of interesting, and interested, men for short-term mating. Some women are attracted to "bad boys," men who are "fickle, frivolous, opportunistic, hardheaded, handsome, confident and conceited."[39] These women prefer larger pupils in the eyes of men.

The sclera—the white of the eye—affects attraction. No other primates have white scleras. Their scleras are dark, hiding their direction of gaze from predators, and from members of their own species—for whom a stare can be a threat.[40] The white sclera of the human eye advertises gaze direction, making it a tool for social communication. It also advertises emotion and health. The sclera is covered by the conjunctiva, a thin membrane containing tiny blood vessels. Certain emotions, such as fear and sadness, and certain pathologies, such as allergies and conjunctivitis, cause these vessels to dilate, making the sclera red. This is not lost on our genes. Photos of faces in which the whites of the eyes are artificially reddened look emotional and less attractive.[41] Liver disease and aging can add a yellow cast to the sclera. Whitening the sclera makes a face more attractive.[42]

The sclera in infants is thin, allowing the choroid below to give the white sclera a bluish cast.[43] As we age, the sclera thickens and this cast disappears. So bluish scleras are correlated with youth. Because men prefer youth in women, and women prefer slightly older men, I predicted that men, more than women, prefer bluer scleras in the opposite sex. Sammaknejad tested this prediction. She showed a sequence of faces and had observers use a slider to adjust the hue of their scleras, from bluish to yellowish, until each face looked most attractive. Women adjusted male scleras to be slightly blue, but men, as predicted, adjusted female scleras to be bluer.[44] Once again, a subtle cue to fitness is picked up by our genes. One application is clear. To make your portrait more attractive, don't just whiten your scleras. Add a hint of blue. Women should add a tad more blue than men.

Our eyes, being moist, also sparkle with highlights, which enhance their attraction. Professional photographers know this and use "catch lights" to add highlights on the eyes. Painters know this as well: the eyes of Vermeer's *Girl with a Pearl Earring* sparkle with life; the eyes of the *Mona Lisa* have no sparkle, adding to her enigma. Anime cartoons exaggerate highlights to heighten the attraction of their characters. Filmmakers avoid highlights in the eyes of villains, making them lifeless and nefarious.

Highlights on the eyes reflect from a film of tears, produced by lacrimal glands, that veil the cornea and sclera.[45] This film grows thin and our eyes grow dry as we age or suffer disease, such as Sjögren's syndrome, lupus, rheumatoid arthritis, thyroid disease, and meibomian gland dysfunction. A dry eye reflects less light than one covered with an ample film.[46] So brighter highlights signal youth and health.

Does our feeling of attraction track this signal? Darren Peshek found that indeed it does. Faces with highlights are more attractive than faces with no highlights or dim highlights. But if the highlight in one eye is higher than in the other—suggesting an asymmetry between the eyes—then the face is much less attractive. If you add highlights to your portrait, take care that they are vertically aligned.

Humans are not alone in their attention to highlights in eyes. The owl butterfly, for instance, has fake owl eyes painted on its wings, each eye flourishing a fake highlight. This attention to detail suggests an evolutionary arms race in which fake eyes—in order to scare avian predators—grew ever more realistic as the vision of hungry birds grew ever more discerning. At some point in this race, a mutation—perhaps affecting genes such as the Engrailed, Distal-less, Hedgehog, or Notch genes[47]—daubed a highlight on an eyespot that was lifelike enough to scare off birds, and the mutation caught on. This arms race is oft repeated: many species of butterflies and moths, in their battle to survive, flaunt eyespots with fake highlights.

Fake highlights can also promote love. For females of the African butterfly *Bicyclus anynana*, the highlights on a male's eyespots, if crafted just right, are a turn-on. If his smell is also up to par, they are irresistible.[48] Why are fake sparkles so alluring? A male whose eyespots have the right sparkles is better at scaring off predators and staying alive. A female attracted to him is more likely to have offspring with eyespots that scare off predators. So the genes behind her attraction are more likely to spread. Fake highlights attract love because they avoid war.

Genes have other strategies with eyespots. The large and flamboyant tail of the peacock, for instance, with its spray of hypnotizing eyespots, signals to the peahen that, despite this weighty handicap, he's fit enough to avoid predation, and thus fit enough to warrant her affection.[49] Genes use many schemes to push their way into the next generation. All's fair in love, war, and snatching fitness points.

The eyes of animals on land sparkle with highlights because the index of refraction of light in air differs from its index in the film of tears on the eyes. For creatures in water, this difference of index disappears and with it the sparkle of highlights in their eyes. Some fish—such as the eyespot goby, the ambon damselfish, and the copperband butterfly fish—evolved eyespots as a defense against predators. But their eyespots lack highlights because eyes in water lack highlights. The fitness payoff for fake highlights depends on the context: some if by land, none if by sea.

Your genes ply a variety of strategies to finagle their way into the next generation. It wasn't until 1963 that William Hamilton, then a graduate student in London, discovered that the genes inside your body can also push the genes inside *other* bodies into the next generation. Not just any other bodies, but bodies that contain genes related to your own. You share half of your genes with your siblings and parents, a quarter with your grandkids, and an eighth with your cousins. Hamilton discovered that natural selection permits a strategy to survive if it confers a benefit of fitness to a relative that is greater than its cost of fitness to you. How much greater depends on how related you are. The benefit to your brother or sister must be at least twice the cost to you; the benefit to a grandchild at least four times the cost to you; and the benefit to a cousin at least eight times the cost to you. This broader notion of fitness is called "inclusive fitness" to distinguish it from the notion of "personal fitness," which we have discussed until now.[50] The two notions are not at odds. Inclusive fitness simply recognizes a broader spectrum of strategies by which genes muscle into the next generation.

Inclusive fitness can explain the evolution of some altruistic behaviors, which enhance the fitness of others at a cost to oneself. An example is the alarm call of the Belding's ground squirrel, a native of the northwestern United States, which sits low on the food chain and high on the menu for eagles, weasels, bobcats, badgers, and coyotes.[51] If a wary squirrel detects an eagle, it shrieks an alarm, even if it is exposed and vulnerable. It warns nearby squirrels and risks its own life by calling attention to itself. If nearby squirrels share genes for shrieking an alarm, this strategy lubricates the passage of these genes to the next generation even if, now and then, a sentinel becomes a meal. The genes survive even if, and indeed because, some squirrels are sacrificed; that's a risk the genes are willing to take. There are limits, however, to the altruism of squirrels. When a predator comes by land, rather than by air, a squirrel darts to safety before shrieking.

A gene in you that forfeits you to save your neighbor can survive if it also resides in that neighbor. The chance of coresidence depends on your genetic relatedness. Because we cannot inspect DNA, our genes have evolved strategies that fallibly but adequately estimate relatedness. One strategy assumes that your conspecifics—members of your own species—that are nearby are more related to you than those further away. This is true often enough to shape a useful heuristic: show more altruism toward those you more often see.[52]

Another strategy estimates relatedness from sensory cues. A female Belding's ground squirrel, for instance, relies heavily on scents to estimate relatedness, and favors those who smell more related to her.[53]

Larry Maloney, a professor of psychology at New York University, and Maria dal Martello, a professor of psychology at Padua University in Italy, found that we can estimate kinship between strangers by looking at faces. We glean more information about kinship from the upper half of the face than from the lower. The eyes, in particular, account for one-fifth of our ability.[54] The features of the eyes that influence our estimate of kinship are not yet known.

We have seen in this chapter that features of eyes, such as the limbal ring, can make us attractive and thereby enhance our personal fitness. The eyes, as it happens, also inform us about kinship and thereby enhance inclusive fitness. The eyes may be windows to the soul, but they are certainly windows to what matters most in evolution: fitness, both personal and inclusive.

I focused in this chapter on the beauty of eyes, both for brevity and because we spend more time watching eyes than any other objects. Our genes, of course, estimate fitness using hundreds of other sensory cues, such as height, weight, smell, and quality of voice.[55]

Genes shape male perceptions of female beauty. To be clear, this fact does not justify sexism, patriarchy, or oppression of women. The discovery that genes influence our emotions and behavior does not justify an oppressive status quo any more than the discovery that genes influence cancer justifies cancer. To the contrary, the advance of evolutionary psychology provides tools to understand and prevent oppression, just as the advance of molecular biology provides tools to understand and treat cancer.

Evolutionary psychology reveals that our perception of beauty is an estimate of reproductive potential. This does not entail that we have sex only to procreate. Exaptation, in which a trait evolved for one function can co-opt a new function, is commonplace in nature. We use sex to procreate, but also to bond, play, heal, and enjoy pleasure.

With these provisos, our study of beauty is just the background we need to grapple with our central question: Do we perceive reality as it is? We will find a counterintuitive answer. If our senses evolved and were shaped by natural selection, then spacetime and physical objects, like beauty, reside in the eye of the beholder. They inform us about fitness—not about truth or objective reality.

CHAPTER THREE

Reality

Capers of the Unseen Sun

> "Evolutionarily speaking, visual perception is useful only if it is reasonably accurate. . . . Indeed, vision is useful precisely because it is so accurate. By and large, what you see is what you get. When this is true, we have what is called veridical perception . . . perception that is consistent with the actual state of affairs in the environment. This is almost always the case with vision."
>
> —STEPHEN PALMER, *VISION SCIENCE*

"I don't see why you pick on neurons," Francis Crick wrote on April 13, 1994. "Surely you believe the sun existed before there was anyone to perceive it. So why should neurons be different?" A few weeks earlier, Crick had kindly sent me a signed copy of his new book *The Astonishing Hypothesis*. I read it and sent him a letter, on March 22, thanking him for the book. I also raised a question about its hypothesis:

> Perhaps you could help me escape what seems a paradox. I agree wholeheartedly with you that "seeing is an active, constructive process," that what we see "is a symbolic interpretation of the world," and that "in fact we have no direct knowledge of objects in the world." Indeed I think perception to be like science: a process of constructing

theories given the available evidence. We see the theories we believe.
As you say, "seeing is believing."

On these points, Crick and I agreed. But they conflict with common sense, and so they warrant some discussion. Most of us don't claim to know exactly how seeing works. But if pressed, we may speculate that it's much like a video camera. There is, we believe, a real 3D world that exists even when no one looks, and it contains real objects such as red apples and misty waterfalls. When we look, we simply shoot a video of this world. There's really not much to it, and most of the time it works quite well—our video shots are accurate.

But common sense is in for a surprise. Neuroscientists assure us that each time we open our eyes, billions of neurons and trillions of synapses spring into action. Roughly one-third of the brain's cortex, one-third of our most advanced computing power, is engaged in vision—which is not what you may expect if seeing is just a matter of shooting videos. Cameras, after all, were filming long before the era of the computer. So what in the world is the brain computing when we look, and why?

The standard reply by neuroscientists is that the brain is constructing, in real time, our perceptions of objects such as apples and waterfalls.[1] It constructs them because the eye itself does not see apples and waterfalls; instead, it has about 130 million photoreceptors, and each of them sees just one thing: how many photons of light it just captured. So the photoreceptors are bean counters for photons, and issue boring reports, something like this: Photoreceptor #1: twenty photons; Photoreceptor #2: three photons; . . . Photoreceptor #130,000,000: six photons. There are, at the photoreceptors of the eye, no luscious apples and no dazzling waterfalls. There is just a stupefying array of numbers, with no obvious meaning. To endow this hill of beans with meaning, to understand what these lifeless numbers say about a living world, is such a daunting task that billions of neurons, including many millions within the eye itself, are conscripted into service. It's not like translating Greek to English. It's more like detective work: the numbers are cryptic clues, and the brain must sleuth like Sherlock. Or it's like theoretical physics: the numbers are experimental data and the brain must pull an Einstein. With clever detective work and theorizing, your brain interprets a jumble of numbers as a coherent world, and that interpretation is what you see—the best theory your brain could muster.

Which is why Crick claimed, and I agreed, that "seeing is an active, constructive process," that what we see "is a symbolic interpretation of the world," that "in fact we have no direct knowledge of objects in the world," and that seeing is believing your best theory.

But then I set up my paradox. If we construct everything we see, and if we see neurons, then we construct neurons. But what we construct doesn't exist until we construct it (too bad; it would be much cheaper to move into my dream mansion before constructing it). So neurons don't exist until we construct them.

But this conclusion, I wrote in that March 22 letter, "contradicts, it would seem, the astonishing hypothesis, viz., that neurons exist prior to and are, somehow, causally responsible for, our perceptions."

I didn't expect that Crick would buy my argument. But I was interested to hear why. He wrote back on March 25, 1994: "It is a reasonable hypothesis that a real world consists of which we only have limited knowledge and that neurons existed prior to anyone observing them *as neurons*." (The emphasis is by Crick, which he indicated by underlining.)

Crick argued, and most neuroscientists would agree, that it's reasonable to assume that neurons exist prior to anyone perceiving them as neurons. But I wanted to better understand his thoughts on the relation between perception and reality. So in a letter on April 11, 1994, I pressed further. "We can, as you say, *hypothesize*, that neurons exist in the world prior to any representations of them. But this hypothesis, though reasonable, is untestable. How shall we, in principle, falsify it?"

This prompted Crick to reply on April 13: "I don't see why you pick on neurons. Surely you believe the sun existed before there was anyone to perceive it. So why should neurons be different?" But then, as I had hoped, he shared his thoughts on perception and reality. "It seems to me, following

Kant, one has to distinguish between the thing-in-itself (the sun in the above example), which is essentially unknowable, and the 'idea-of-the-thing,' which is what our brains construct. Then the argument becomes what are perceived are symbolic constructions. The sun-in-itself can be the subject of perception. Our idea-of-the-sun is a symbolic construction. The idea-of-the-sun does not exist prior to its construction—but the sun-in-itself did!"

Fair enough. Crick rejected, and so do I, metaphysical solipsism, which says that I and my experience are all that exist. According to this solipsism, if I see you then you exist, but only as my experience. When I close my eyes, you cease to exist. I reside in a universe of my own making, a universe of my experiences. I am alone. I cannot join a Society of Solipsists or wonder, without irony, why more people aren't solipsists.

Crick embraced metaphysical realism. The sun-in-itself exists even when no one looks. I only construct my perception of that sun—my idea-of-the-sun.

Most of us are metaphysical realists. It seems to be a view that comes naturally. Suppose, as we discussed in chapter one, that you open your eyes and have an experience that you describe as a red tomato a meter away. Then you close your eyes and your experience changes, to a gray field. Is it still true, while you see gray, that there is a red tomato a meter away? Most of us would say yes. Now this tomato that we believe exists, even when no one looks, is what Crick would call the "tomato-in-itself." It is not the same as your experience of a tomato (or, as philosophers helpfully put it, "your experience *as of* a tomato"), your "idea-of the-tomato."

Crick said in his letter that the thing-in-itself—the tomato-in-itself or the neuron-in-itself—"is essentially unknowable." But most of us believe otherwise. We believe, for instance, that the tomato-in-itself is, like our experience, red and tomato-shaped and a meter away. We believe that experience accurately depicts the thing-in-itself.

I suspected that Crick also believed this. He believed that our idea-of-the-neuron accurately depicts the neuron-in-itself. The 3D shape of a neuron that a neuroscientist experiences when she looks through a microscope tells her the true shape of a neuron-in-itself. The clicks she hears from a microelectrode tell her the true activity of a neuron-in-itself. In his book, Crick said, "The Astonishing Hypothesis is that 'You,' your joys and your sorrows, your memories and your ambitions, your sense of personal identity and free will, are in fact no more than the behavior of a vast assembly of nerve cells and their associated molecules. . . . 'You're nothing but a pack of neurons.'" Crick clearly meant a pack of neurons-in-themselves, not a pack of ideas-of-neurons.

So I wrote him another letter, on May 2, 1994, asking for his thoughts about this central issue.

"The Astonishing Hypothesis is still untestable. For only the idea-of-neuron is observed in experiments, not the neuron-in-itself. And the only way to bridge this gap, so far as I can see, is to hypothesize that the neuron-in-itself is, in important ways, similar to our idea-of-neuron. (These remarks, if correct, hold also for the sun-in-itself and so on.) Let's call this the *Bridge Hypothesis* . . .)

"In short, I think the Astonishing Hypothesis, even in its revised form, is untestable. Or rather, it is testable only if one assumes the Bridge Hypothesis which, since it asserts a relationship between the perceived and the unperceivable, is itself untestable and dubious. . . . The thing-in-itself is ontological baggage, not useful for the scientific enterprise."

I didn't buy the bit about baggage, and I figured Crick wouldn't buy it either, but I wanted to hear his thoughts.

Crick responded on May 4, 1994: "I don't think it sensible to discard the "thing-in-itself," as the idea is of some use in warning us about what we cannot know. It is, however, a *hypothesis* that we can *usefully* talk in this way, but it is the standard hypothesis underlying all science, even (I think) quantum mechanics. The problem only becomes acute when we discuss qualia."

The term *qualia* is sometimes used by philosophers to refer to subjective, conscious experiences—what it's like to see the redness of red or smell the aroma of coffee. I will avoid this term because it often triggers debates about its precise definition. I will instead refer to conscious experiences.

Crick continued. "In fact, our present tentative view of the way the brain works would suggest that some aspects of qualia *cannot* be communicated. The problem, rather, is to explain why qualia exist at all. The party line is that we should try to find out the NCC (the Neural Correlate of Consciousness) before we worry too much about this aspect of qualia."

Crick was pragmatic about the thing-in-itself: it is a *hypothesis* that we can *usefully* talk this way (he underlined "hypothesis" and "usefully"). He was frank about the problem of conscious experiences. Their very existence was, he thought, too hard to explain at the time. In his quest to understand DNA, Crick was famously influenced by Schrodinger's thoughts about genes in the book *What is Life?* Apparently, Crick was also influenced by Schrodinger's thoughts, in that same book, about conscious experiences: "The sensation of color cannot be accounted for by the physicist's objective picture of light-waves. Could the physiologist account for it, if he had fuller knowledge than he has of the processes in the retina and the nervous processes set up by them in the optical nerve bundles and in the brain? I do not think so."

Crick assumed, however, that the thing-in-itself can be described using the vocabulary of our ideas-of-things, of objects moving in space and time. Heat-in-itself, for instance, is molecular motion in space and time; a neuron-in-itself is an object with a shape and activity that evolves in space and time. He assumed that our ideas-of-things truly describe the thing-in-itself, so that the same vocabulary describes both. I rejected this assumption as implausible. But Crick thought it applied even to objects, space, and time.

Crick was supported in his view by a young neuroscientist, David Marr, who revolutionized our understanding of vision in the late 1970s and early 1980s. Crick met Marr in England. Crick then moved to the Salk Institute in San Diego, and Marr moved to MIT. In April of 1979, Marr and his colleague Tomaso Poggio spent a month with Crick at the Salk, discussing visual neuroscience.

Marr claimed that our perceptions normally match reality, that our ideas-of-things correctly describe the things-in-themselves. As he put it in his 1982 book *Vision*: "usually our perceptual processing does run correctly (it delivers a true description of what is there)." He believed that this match between perception and reality was the result of a long process of evolution: "We . . . very definitely do compute explicit properties of the real visible surfaces out there, and one interesting aspect of the evolution of visual systems is the gradual movement toward the difficult task of representing progressively more objective aspects of the visual world."

The human visual system, Marr argued, evolved its ideas-of-things to match the true structure of the things-in-themselves, although the match is not always perfect: "usually our perceptual processing does run correctly (it delivers a true description of what is there), but although evolution has seen to it that our processing allows for many changes (like inconstant illumination), the perturbation due to the refraction of light by water is not one of them." But Marr concluded that natural selection had, on balance, shaped our perceptions to match reality: "The payoff is more flexibility; the price, the complexity of the analysis and hence the time and size of brain required for it."

Crick argued that the thing-in-itself is a useful hypothesis. Marr argued further, on evolutionary grounds, that our perceptions, our ideas-of-things, depict reality, the thing-in-itself, with accuracy. In my 1994 exchange with Crick, I had no counter to Marr's argument from evolution for the Bridge Hypothesis.

Indeed, my thoughts on perception and reality were shaped by Marr. I first encountered his ideas in a graduate class on Artificial Intelligence at UCLA in the 1977–78 academic year. I was a senior, working toward a Bachelor of Arts in Quantitative Psychology, but Professor Edward Carterette kindly allowed me into his graduate class. One paper we discussed was by Marr. I found it electrifying in style and content. Marr built models of vision that were precise enough to be programmed into a computer. If the computer was then linked to video cameras, these programs could analyze the images received from the cameras, and infer important features of the nearby environment, such as its 3D structure. Marr's goal was clear: create precise models of human vision and use them to build computers and robots that see.

I was hooked. Where was this guy, and how could I work with him? I was surprised to learn that Marr was in the Psychology Department at MIT. Psychology at MIT? I thought of MIT as a bastion of math and hard science, not psychology. I later learned that Marr was also in the Artificial Intelligence Laboratory. I decided to apply to MIT to be his student. The Cold War was at full fever, and I worked my way through UCLA as a cold warrior, employed by Hughes Aircraft to write flight simulators and cockpit displays for fighter jets, such as the F-14, in the machine code of a microprocessor called the AN/UYK-30. I graduated from UCLA in June of 1978, continued at Hughes for another year, and entered MIT in the fall of 1979 as Marr's graduate student.

I soon learned that Marr had leukemia. He died fourteen months later, in November of 1980, at the age of thirty-five. But those fourteen months exceeded my expectations. Marr inspired in person as he did in print. He was the center of gravity for a community of eager students and brilliant colleagues. Discussions were lively, multidisciplinary, and game-changing.

There were ups. Marr went into remission and married Lucia Vaina. There were downs. Jeremy, a grad student in psychology, completed his PhD that spring and the next day took his life—the rumor was cyanide. All of the grad students were dazed. Days later, as I walked by Marr's office on the eighth floor of the Artificial Intelligence Lab, he waved me in. "If you ever feel like ending your life, come see me first. Life is worth living."

Marr soon came to lab meetings visibly weakened, with a handkerchief over his nose and mouth. Then, tragically, not at all. Whitman Richards, a brilliant psychophysicist and advocate of Marr's ideas, was my coadvisor while Marr was alive, became my sole advisor after his death, and remained a dear friend until his own death in 2016.

I completed my PhD in the spring of 1983, and in the fall joined the Department of Cognitive Sciences at UC Irvine. By 1986, I doubted Marr's claim that we evolved "to see a true description of what is there." I also doubted that the language of our perceptions—the language of space, time, shapes, colors, textures, smells, tastes, and so on—can frame a true description of what is there. It is simply the wrong language. But I was unable, in 1994, to offer Crick a good argument against Marr's claim.

Indeed, there is, to the contrary, a stock argument in its favor: those of our predecessors who saw reality more accurately had a competitive advantage over those who saw it less accurately. They were more likely to pass on their genes that coded for more accurate perceptions. We are the offspring of those who, in each generation, saw more accurately. So we can be confident that, after thousands of such generations, we see reality as it is. Not all of reality, of course. Just the parts that matter for survival in our niche. As Bill Geisler and Randy Diehl put it: "In general, (perceptual) estimates that are nearer the truth have greater utility than those that are wide of the mark."[2] Thus, "In general, it is true that much of human perception is veridical [accurate] under natural conditions."[3]

The evolutionary theorist Robert Trivers, whose insights into evolution transformed our understanding of social relations, makes a similar argument. "Our sense organs have evolved to give us a marvelously detailed and accurate view of the outside world . . . our sensory systems are organized to give us a detailed and accurate view of reality, exactly as we would expect if truth about the outside world helps us to navigate it more effectively."[4]

Vision scientists disagree on many technical issues, such as the role of action and embodiment in perception, and whether perception involves construction, inferences, computations, and internal representations. But they do agree on this: the language of our perceptions is suitable to describe what exists when no one looks; and, in the normal case, our perceptions get it right.

For instance, in his textbook *Vision Science*, Stephen Palmer tells students of perception that "Evolutionarily speaking, visual perception is useful only if it is reasonably accurate." The idea is that perceptions that are truer, that better match the state of the objective world, are thereby fitter. So natural selection shapes our perceptions to be truer.

Most perceptual theorists propose that the brain creates internal representations of the outside world, and that these internal representations are responsible for our perceptual experiences. They claim that our experiences are veridical, meaning that the structure of these internal representations, and therefore of our experiences, matches the structure of the objective world.

Alva Noë and Kevin O'Regan tell us, "Perceivers are right to take themselves to have access to environmental detail."[5] Noë and O'Regan agree that the brain creates internal representations of the outside world, but claim that these internal representations are not responsible for our experiences. They propose instead that our perceptual experiences arise from our active exploration of the objective world, and our discovery, in this process, of contingencies between our actions and perceptions. But they agree that this process results in perceptual experiences that are veridical.

Zygmunt Pizlo and his colleagues tell us, "veridicality is an essential characteristic of perception and cognition. It is absolutely essential. *Perception and cognition without veridicality would be like physics without the conservation laws.*"[6] The emphasis is theirs. Pizlo argues that our perceptions are veridical because evolution has shaped our sensory systems to perceive real symmetries in the outside world.

Some researchers, such as Jack Loomis, agree that there are similarities between our perceptions and objective reality, but contend that our perceptions can have systematic errors, especially of perceived shape.[7] These researchers assume, however, that the language of our perceptions is the right language to frame true descriptions of what is there.

But despite the consensus of experts, I doubted that natural selection favors perceptions that describe reality. More deeply, I doubted that selection favors perceptions that could even frame true descriptions of reality. It's not that on occasion a perception exaggerates, underestimates, or otherwise goes awry, it's that the lexicon of our perceptions, including space, time, and objects, is powerless to describe reality.

I found an argument for doubt from Marr himself, in his book *Vision*, an argument he aimed at simpler organisms, such as flies and frogs. "Visual systems like the fly's . . . are not very complicated; very little objective information about the world is obtained. The information is all very subjective." He argued that "it is extremely unlikely that the fly has any explicit representation of the visual world around him—no true conception of a surface, for example." But he thought that, despite its failure to represent the world, the fly could still survive because it can, for instance, "chase its mate with sufficiently frequent success."[8]

Then Marr explained how a simple system that "does not really represent the visual world about it" may nevertheless evolve. "One reason for this simplicity must be that these facts provide the fly with sufficient information for it to survive."[9]

Marr argued that natural selection can favor simple, subjective perceptions, that don't represent objective reality, if they do guide adaptive action. This raises the question: When does natural selection favor veridical perceptions over subjective perceptions? Marr answered: when organisms get more complex. Humans, he claimed, have veridical perceptions, and simple flies do not. But is this correct?

Perhaps not. The cognitive scientist Steven Pinker has explained why natural selection may not favor veridical perceptions. My last year at MIT as a graduate student was Pinker's first year there as an assistant professor. I had the pleasure of taking one of his classes and becoming dear friends. It was obvious then that, with his creativity, incisive logic, and encyclopedic mastery of the literature, he would make stellar contributions to the cognitive sciences, as in fact he has. His 1997 book, *How the Mind Works,* focused my attention on evolutionary psychology.[10] Before I read his book, I knew about evolutionary psychology and the groundbreaking work of Leda Cosmides and John Tooby. Indeed, I had tried and failed to persuade my department, in 1991, to offer Leda a faculty position—evolutionary psychology was, and still is, controversial. It has been accused, for instance, of lacking hypotheses that are testable, justifying unsavory moral and political ideas, and claiming that human behavior is determined by genes, with little influence from the environment. These accusations are misguided.

Pinker's book persuaded me to study perception as a product of natural selection. He makes a surprising claim: "Our minds evolved by natural selection to solve problems that were life-and-death matters to our ancestors, not to commune with correctness." This observation is central. Our minds were shaped by natural selection to solve life-and-death problems. Full stop. They were not shaped to commune with correctness. Whether our beliefs and perceptions happen to be true is a question that requires careful study.

In his critique of *How the Mind Works*, Jerry Fodor argued that no such study is needed, because nothing in science "shows, or even suggests, that the proper function of cognition is other than the fixation of true beliefs."[11]

In reply, Pinker offered several reasons why beliefs may evolve to be false.[12] For instance, computing the truth is costly in time and energy, and so we often use heuristics that risk being false or out of date. Pinker conceded, however, that "We do have some reliable notions about the distribution of middle-sized objects around us."[13]

What about those middle-sized objects around us—tables, trees, and tomatoes? When we see them, it feels like we see the truth. Most vision scientists concur: if I see a tomato and then close my eyes, the tomato is still there.

But could we be wrong? Is it possible that there is no tomato if no one looks? No space and time? No neurons? No neural activity to cause, or be, our conscious experiences? Is it possible that we do not see reality as it is?

Stephen Hawking and Leonard Mlodinow argue for a model-dependent realism: "According to model-dependent realism, it is pointless to ask whether a model is real, only whether it agrees with observation. If there are two models that both agree with observation . . . then one cannot say that one is more real than another."[14]

Hawking and Mlodinow then ask: "How do I know that a table still exists if I go out of the room and can't see it?. . . . One could have a model in which the table disappears when I leave the room and reappears in the same position when I come back, but that would be awkward. . . . The model in which the table stays put is much simpler and agrees with observation."[15]

Indeed, if two models agree with observation, then prefer the simpler. But the model in which the neuron stays put has so far, and despite valiant efforts by talented neuroscientists, failed to explain the origin, nature, and data of conscious experience: no theory that starts with neurons and neural activity can account for observations about conscious experiences and their correlations with neural activity. Perhaps the model in which the neuron stays put is an impediment to our progress in understanding the origin of consciousness.

Philosophers have, for centuries, debated the puzzle of perception and reality. Can we transform this philosophical puzzle into a precise scientific question? Can Darwin's theory of natural selection provide a definitive answer?

In 2007, I decided to try. It was time to see if neurons stay put, or if we should pick on them.

CHAPTER FOUR

Sensory

Fitness Beats Truth

"Little did I realize that in a few years I would encounter an idea—Darwin's idea—bearing an unmistakable likeness to universal acid: it eats through just about every traditional concept, and leaves in its wake a revolutionized world-view, with most of the old landmarks still recognizable, but transformed in fundamental ways."

—DANIEL DENNETT, *DARWIN'S DANGEROUS IDEA*

"If you ask me what my ambition would be, it would be that everybody would understand what an extraordinary, remarkable thing it is that they exist, in a world which would otherwise just be plain physics. The key to the process is self-replication."

—RICHARD DAWKINS, IN JOHN BROCKMAN'S *LIFE*

Most of us assume that we normally see reality as it is; if you see an apple, that's because there really is an apple. Many scientists assume that we have evolution to thank for this—accurate perceptions enhance our fitness, so natural selection favors them, especially in species like *Homo sapiens* with bigger brains. Most neuroscientists and experts in perception agree. They sometimes say that our perceptions recover, or reconstruct, the shapes and colors of real objects; many don't bother to mention it because it's just too obvious.

But are they right? Does natural selection favor true perceptions? Is it possible that we did not evolve to see truly—that our perceptions of space, time, and objects do not reveal reality as it is? That a peach does not exist when no one looks? Can the theory of evolution transform this stale philosophical chestnut into a crisp scientific claim?

Some say no: the notion that a peach isn't there when no one looks is irremediably unscientific. After all, what observation could possibly tell us what happens when no one observes? None. It's a self-contradiction. This half-baked proposal can't be tested by an experiment, so it's metaphysics, not science.

This rejoinder misses a point of logic and a matter of fact. First, logic: if we can't test the claim that a peach does not exist when no one looks, then we can't test the opposite and widely held claim that it does exist. Both claims posit what happens when no one observes. If one is not science, then neither is the other. Nor is the claim that the sun exists when no one looks, that the big bang happened over thirteen billion years ago, and other such claims routinely made in science.

Now the matter of fact: observation *can* test a claim about what happens when no one looks. One can be pardoned for not realizing this. Even the brilliant physicist Wolfgang Pauli missed it, and likened such claims to "the ancient question of how many angels are able to sit on the point of a needle."[1] But in 1964, the physicist John Bell proved him wrong: there are experiments that can test such claims—for instance, the claim that an electron has no spin when no one looks.[2] Bell's experiments have been run, in many variations, with consistent results. Bell's Theorem transported such claims from the realm of angels to the beat of science. We will discuss how in chapter six.

Thus these claims are in the purview of science. But are they in the purview of evolution? Can we ask, precisely, if natural selection favors true perceptions? Can we expect the theory of evolution to render a verdict?

Some argue that it cannot: perceptions that are true must also enhance fitness. Truth and fitness, they claim, are not rival strategies, but rather the same strategy, seen from different perspectives.[3] Thus evolution cannot render an impartial verdict.

This argument fails because it forgets a simple point about fitness: according to standard accounts of evolution, although fitness payoffs depend on the true state of the world, they also depend on the organism, its state, its action, and its competition. Feces, for instance, offer big payoffs for hungry flies, but not for hungry humans. A hydrothermal vent, belching hydrogen sulfide at 80°C into water a few kilometers deep, offers big payoffs for the Pompeii worm *Alvinella pompejana*, but hideous death to all but a handful of extremophiles. The distinction between a state of the world (say, a pile of feces) and the fitness payoffs it offers to an organism (say, a fly or a man) is essential in evolution.

According to standard accounts of evolution, payoffs can vary wildly while the true state of the world remains fixed. It follows that seeing truth and seeing fitness are two distinct strategies of perception, not one strategy seen in different lights. The two strategies can compete. One may dominate and the other go extinct. So it is a central question, not a conceptual mistake, to ask: Does natural selection favor perceptions tuned to truth or to fitness?

Some argue that the theory of evolution cannot address this question, because the answer may refute the theory. Evolution assumes that there are physical objects in space and time, such as DNA, RNA, chromosomes, ribosomes, proteins, organisms, and resources. It could not, without refuting itself, conclude that natural selection drives true perceptions to extinction. For then the very language of space, time, and physical objects would be the wrong language to describe objective reality. Our scientific observations of physical objects in spacetime, such as DNA, RNA, and proteins, would not be veridical descriptions of objective reality, even if these observations use advanced technologies, such as X-ray diffractometers and electron microscopes. The theory of evolution would refute itself by discrediting its own key assumptions—the logical equivalent of shooting itself in the foot.

It is true that evolution by natural selection, as Darwin himself described it, assumes the existence of "organic beings." But Darwin's own summary of his theory hints that the real work is done by an abstract algorithm—variation, heredity, and selection. "But if variations useful to any organic being do occur, assuredly individuals thus characterized will have the best chance of being preserved in the struggle for life; and from the strong principle of inheritance they will tend to produce offspring similarly characterized. This principle of preservation, I have called, for the sake of brevity, Natural Selection."[4]

This algorithm of variation, heredity, and selection applies to organic beings but, as Darwin recognized, it also applies more broadly and to more abstract entities, such as languages. "Languages, like organic beings, can be classed in groups under groups; and they can be classed either naturally according to descent, or artificially by other characters. Dominant languages and dialects spread widely, and lead to the gradual extinction of other tongues."[5]

Thomas Huxley realized that Darwin's algorithm applied to the success of scientific theories. "The struggle for existence holds as much in the intellectual as in the physical world. A theory is a species of thinking, and its right to exist is coextensive with its power of resisting extinction by its rivals."[6] Richard Dawkins proposed that Darwin's algorithm applies to "memes," units of cultural transmission such as "tunes, ideas, catch-phrases, clothes fashions, ways of making pots or of building arches."[7] Memes can pass from person to person, and can be altered in the process. "This land is your land" was first a meme in the mind of Woody Guthrie, but it proliferated, with variations, into the minds of Peter, Paul, and Mary, Bob Dylan, and others, competing successfully against many songs for the limited time, interest, attention, and memory of human minds. Many a song that we've never heard was once a meme in someone's mind but had less success at replication.

Darwin's algorithm has been applied to fields such as economics, psychology, and anthropology. The physicist Lee Smolin applied it to the largest scale of all—cosmology—proposing that each black hole is a new universe, and that a universe more likely to produce black holes is more likely to produce more universes.[8] Our universe has the properties that it does—such as the strengths of the weak, strong, gravitational, and electromagnetic forces—because they are conducive to creating black holes and, through them, new universes. Universes quite different from ours are less likely to produce black holes, and thus less likely to reproduce.

The insight that Darwin's algorithm applies not just to the evolution of organic beings but also, with some changes, to a variety of other domains, is called *universal Darwinism*.[9] (Richard Dawkins coined the term when arguing that Darwin's algorithm governs the evolution of life not just on earth but anywhere in the universe.) Universal Darwinism, unlike the modern theory of biological evolution, does not assume the existence of physical objects in space and time. It is an abstract algorithm, with no commitment to substrates that implement it.

Universal Darwinism can, without risk of refuting itself, address our key question: Does natural selection favor true perceptions? If the answer happens to be "No," then it hasn't shot itself in the foot. The uncanny power of universal Darwinism has been likened by the philosopher Dan Dennett to a universal acid: "There is no denying, at this point, that Darwin's idea is a universal solvent, capable of cutting right to the heart of everything in sight. The question is: what does it leave behind? I have tried to show that once it passes through everything, we are left with stronger, sounder versions of our most important ideas. Some of the traditional details perish, and some of these are losses to be regretted, but good riddance to the rest of them. What remains is more than enough to build on."[10]

We can apply Darwin's acid to our belief in true perception. We will find that this belief perishes: natural selection drives true perceptions to swift extinction. The very language of our perceptions—space, time, and physical objects—is simply the wrong language to describe objective reality. Darwin's acid dissolves the claim that objective reality consists of spacetime and objects—such as DNA, chromosomes, and organisms. What remains is universal Darwinism, which we can employ even after we jettison spacetime and objects.

How do we apply the acid? In particular, how can we coax Darwin's abstract algorithm to give a concrete answer? Fortunately, the theoretical biologists John Maynard Smith and George Price found a way in 1973—evolutionary game theory.[11] The basic idea is best understood by example.

Camaraderie is not the strong suit of the scorpion *Paruroctonus mesaensis*.[12] When one scorpion detects vibrations that betray the movement of a rival, it pivots and clutches the intruder with its two claws. The intruder immediately snaps its tail trying to sting the attacker, whereupon each scorpion grabs the tail of the other with one claw, and some part of its body with the other. No-holds-barred wrestling ensues until one scorpion sneaks its sting through a chink in the armor of the other, and delivers a lethal injection. It then dines on its conquest, liquifying it with digestive juices and slurping the refreshment. This catch of the day is no rare repast. Cannibalism furnishes 10 percent of a scorpion's menu and, the females agree, is great after sex.

In the battle for mates and territories, some animals—including lions, chimps, humans, and scorpions—kill their rivals. But others battle with ritual or restraint: combatants obey rules of engagement.[13] Some snakes, for instance, sheathe their fangs, and wrestle. Mule deer fight antler to antler, often intensely, and take no cheap shots elsewhere on the body. Why would belligerents obey rules in such contests? Why this glaring exception to "nature red in tooth and claw" and "all is fair in love and war"?

We find an answer in a simple game in which players compete for resources, using one of two strategies: *hawk* or *dove*. A hawk always escalates a conflict. A dove backs down if its opponent escalates.[14] All hawks and doves are equally strong. If the payoff for winning a contest is, say, twenty points, but the cost of injury is, say, eighty points, what will happen? If two hawks compete, neither backs down until one is hurt and the other wins. Because they have equal strength, each hawk wins half the time and gets twenty points for each win. But each hawk gets hurt half the time and loses eighty points for each injury. So when hawks fight each other they lose, on average, thirty points. Their fitness suffers. If two doves compete, each wins half the time and gets twenty points. No dove is hurt. So each dove wins, on average, ten points. Their fitness improves. If a hawk meets a dove, then the hawk wins and no one is hurt. The hawk gets twenty points for a win. The dove gets nothing. Fitness improves for the hawk, but not for the dove.

We can summarize this game in a matrix, shown in Figure 2, which displays the expected payoff to the strategy on the row when it competes with the strategy on the column. So, for instance, the expected payoff for a hawk when it meets a dove is twenty, and the expected payoff for a dove when it meets a hawk is zero.

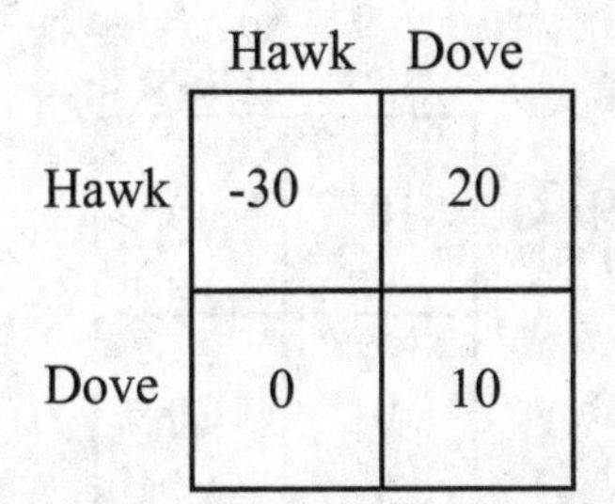

	Hawk	Dove
Hawk	-30	20
Dove	0	10

Fig. 2: Expected payoffs in a hawk-dove game. A hawk, for instance, loses 30 points if it meets another hawk, but gains 20 points if it meets a dove. © DONALD HOFFMAN

Given these payoffs, what strategy is favored by natural selection? The answer depends on the proportion of hawks and doves. Suppose everyone is a hawk. Then everyone loses, on average, thirty points in each competition—a fast track to extinction. Suppose everyone is a dove. Then everyone gains, on average, ten points in each competition—a fast track to greater fitness.

But there is a catch. If everyone is a dove and one hawk shows up, then that hawk has a heyday. It racks up twenty points each time it competes with a dove. This is more than double the points reaped by doves (who get, on average, ten points in contests with other doves and no points in contests with hawks). More fitness points mean more offspring. So this hawk begets more hawks. But the hawk's fun must stop somewhere because, as we saw, if all players are hawks then each loses thirty points on average—the game implodes in extinction.

When does the population of hawks stop growing? When hawks are a quarter of the players. If more than one-quarter are hawks, then hawks earn fewer points than doves. If less than one-quarter of the players are hawks, then hawks earn more points than doves. So, in the long run, one-quarter of the players end up being hawks.

In this example, a win gets twenty points and an injury loses eighty. Change these numbers to forty and sixty. Then the expected payoffs are as shown in Figure 3. Now two-thirds of the players end up being hawks.

	Hawk	Dove
Hawk	-10	40
Dove	0	20

Fig. 3: Expected payoffs in a second hawk-dove game. A hawk now loses 10 points if it meets another hawk, but gains 40 points if it meets a dove. © DONALD HOFFMAN

Fitness depends on payoffs and on how many players adopt each strategy. If everyone is a dove, then it's more fit to be a hawk. If everyone is a hawk, then it's more fit to be a dove. The force of natural selection depends on the frequency of each strategy.[15]

This is a key point. Fitness is no mirror of the world. Instead, fitness depends in complex ways on the state of the world, the state of the organism, and the frequencies of strategies.

If two strategies compete, the dynamics of evolution can be complex. We saw that hawks and doves can coexist. But there are other possibilities. One strategy might always drive the other to extinction—*domination*. Or each strategy might have some chance to drive the other to extinction—*bistability*. Or both strategies might always be equally fit—*neutrality*.

When three strategies compete, the dynamics of evolution allows cycles, as in the classic children's game of Rock-Paper-Scissors: scissors beats paper, which beats rock, which beats scissors.[16] When four or more strategies compete, the dynamics of evolution can include chaos, in which a tiny perturbation now makes unpredictable changes down the road.[17] This is also known as "the butterfly effect"—the flap of the wings of a butterfly here (a tiny perturbation) might trigger a tornado somewhere else (an unpredictable consequence).

All of this can be studied with the theory of evolutionary games. It is a powerful theory. It has the right tools to study our question: Does natural selection favor veridical perceptions?

It gives a clear answer: no.

This is spelled out in the *Fitness-Beats-Truth* (FBT) Theorem, which I conjectured and Chetan Prakash proved.[18] Consider two sensory strategies, each capable of *N* distinct perceptions in an objective reality having *N* states: *Truth* sees the structure of objective reality as best as possible; *Fitness* sees none of objective reality, but is tuned to the relevant fitness payoffs—payoffs that depend on objective reality, but also on the organism, its state, and its action.

> **FBT THEOREM:** *Fitness* drives *Truth* to extinction with probability at least $(N-3)/(N-1)$.

Here's what it means. Consider an eye with ten photoreceptors, each having two states. The FBT Theorem says the chance that this eye sees reality is at most two in a thousand. For twenty photoreceptors, the chance is two in a million; for forty photoreceptors, one in ten billion; for eighty, one in a hundred sextillion. The human eye has one hundred and thirty million photoreceptors. The chance is effectively zero.

Suppose there is an objective reality of some kind. Then the FBT Theorem says that natural selection does not shape us to perceive the structure of that reality. It shapes us to perceive fitness points, and how to get them.

The FBT Theorem has been tested and confirmed in many simulations.[19] They reveal that *Truth* often goes extinct even if *Fitness* is far less complex.

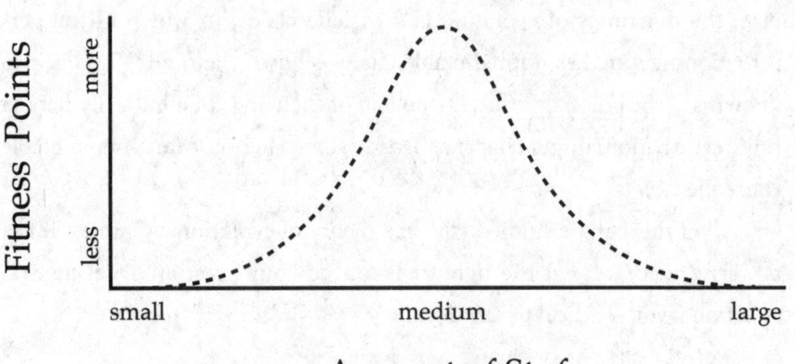

Fig. 4: A fitness function. In this example, small or large amounts of a resource are bad for fitness. Intermediate amounts are best for fitness. © DONALD HOFFMAN

A specific game shows the problem for *Truth*. Consider an artificial world with a creature called a "kritter" that needs a resource called "stuf." If there's too much or too little stuf, then a kritter dies. With the right amount of stuf, a kritter thrives and reproduces. (Stuf affects a kritter as oxygen affects us—too little or too much and we die.) The fitness points that stuf can give to a kritter are plotted in Figure 4. Suppose a kritter has just two perceptions: gray and black. A *Truth* kritter sees as much as it can about the true structure of the world: it sees gray when there's less stuf and black when there's more stuf. A

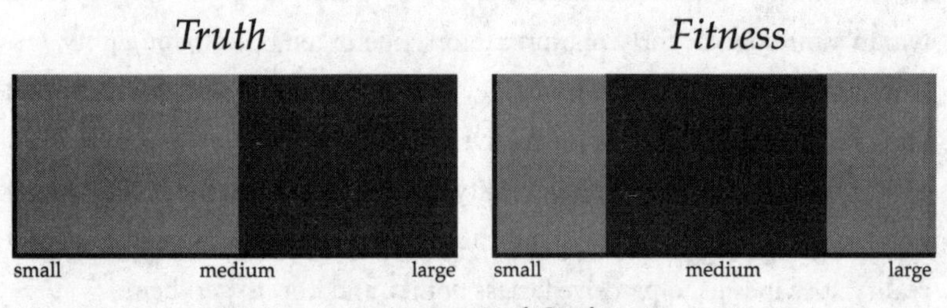

Fig. 5: Seeing truth versus seeing fitness. The shades of gray seen by *Truth* report the amount of a resource but not the fitness payoffs. The shades of gray seen by *Fitness* report the fitness payoffs. © DONALD HOFFMAN

Fitness kritter sees as much as it can about the fitness points available: it sees gray when stuf gives fewer points and black when it gives more. These two strategies, *Truth* and *Fitness*, are shown in Figure 5.

If *Truth* sees gray then it knows there's less stuf. But it knows nothing about the available fitness points. If *Fitness* sees gray then it knows that fewer fitness points are available. But it doesn't know if there is a small or large amount of stuf. Seeing truth hides fitness, and seeing fitness hides truth. Our own senses, for instance, don't perceive oxygen; indeed, we didn't discover oxygen until 1772. Instead, our senses report fitness: we feel a headache if there is insufficient oxygen, and lightheaded if there is too much. Likewise, our senses don't perceive ultraviolet radiation; indeed, we didn't discover this radiation until 1801. Instead, our senses report fitness: we feel sunburn if we receive too much ultraviolet radiation.

If *Fitness* forages for stuf and sees a patch of black, then it knows it is safe to approach. If it sees a patch of gray, then it knows to stay away. But *Truth* has a problem. If *Truth* sees a patch of black, it doesn't know whether it is safe or not. It has the same problem if it sees a patch of gray. So *Truth*, unlike *Fitness*, must risk its life to forage. The truth won't make you free, it will make you extinct.

In Figure 4, as the amount of stuf increases, the number of fitness points first rises and then falls—a bell curve. If, instead, the number of fitness points always increased, then perceptions tuned to fitness would also be tuned to truth, simply because the two are correlated. We know the age of a tree by seeing its rings because the two are correlated—more rings means more years. But if they were not correlated, if some years a tree added rings but other years it erased them, then seeing rings would not tell us the age of the tree.

If fitness payoffs only increase, or only decrease, then perceptions tuned to fitness will also happen to be tuned to truth. So natural selection will happen to favor true perceptions. How likely is this? To answer this question, we count the number of fitness functions that only increase or only decrease. Then we divide by the number of all possible fitness functions. If, for instance, there are six values of stuf and six values of fitness payoffs, then only one fitness function in a hundred allows *Truth* to evolve. If there are twelve values, then only two in a hundred million allow *Truth* to evolve.

In evolution, like football, you win by scoring more points than the competition. Natural selection favors perceptions that assist us in scoring fitness points. If the number of fitness points happens to correlate with a structure in the world, such as the amount of stuf, then evolution will happen to favor *Truth*. But the chance of this is small for simple perceptions and infinitesimal for those more complex.

Stuf has a structure: there can be less or more stuf. But other structures are possible, such as neighborhoods, distances, and symmetries. For each structure we can ask whether fitness points might, by chance, correlate with that structure. And for each we get the same answer: the chance plunges to zero as the world and perception grow more complex. In each case, *Truth* goes extinct when competing with *Fitness*.

Thinkers of stature have claimed the contrary. Marr held that the fly, due to its simplicity, sees no truth, but that mankind, due to its complexity, sees some.[20] He thought that our larger brains permit "the gradual movement toward the difficult task of representing progressively more objective aspects of the visual world."[21] This suits our intuition, but conflicts with the logic of evolution, as revealed by the FBT Theorem.

The notion that our brains are growing in size, and thus in their capacity to see truth, also conflicts with a fact of our evolution: our brains are shrinking.[22] In the last 20,000 years, our brains have shrunk 10 percent—from 1,500 cubic centimeters down to 1,350—a loss of the volume of a tennis ball. Our encephalization quotient, or EQ, which compares our ratio of brain mass to body mass with the average ratio for other mammals, has plunged in an eye blink of evolutionary time. According to the fossil record, this plunge correlates slightly with climate, but heavily with population density and thus, we can presume, with social complexity. This suggests an interesting explanation: the safety net of society eases selection pressures on members; some who wouldn't survive alone, or in small groups, can survive with a larger social net. This possibility, explored with humor in the movie *Idiocracy*, is speculation for now. But the plunge of our EQ is not. If it continues apace it will, within 30,000 years, send our brains back half a million years, to the size of *Homo erectus*. Our brains took the escalator up; they're on the elevator down.

Darwin's idea of natural selection entails the FBT Theorem, which in turn entails that the lexicon of our perceptions—including space, time, shape, hue, saturation, brightness, texture, taste, sound, smell, and motion—cannot describe reality as it is when no one looks. It's not simply that this or that perception is wrong. It's that none of our perceptions, being couched in this language, could possibly be right. The FBT Theorem runs counter to strong intuitions of experts and laymen alike. Dennett was right—Darwin's idea is a "universal acid: it eats through just about every traditional concept, and leaves in its wake a revolutionized world-view, with most of the old landmarks still recognizable, but transformed in fundamental ways."

That revolutionized view leaves in its wake an evolutionary biology that is itself transformed. Still recognizable, after the bath in Darwin's acid, are the landmarks of universal Darwinism: variation, selection, and heredity. But gone from objective reality are physical objects in spacetime, including those central to biology: DNA, RNA, chromosomes, organisms, and resources. This doesn't entail solipsism. *Something* is there in objective reality, and we humans experience its import for our fitness in terms of DNA, RNA, chromosomes, organisms, and resources. But the FBT Theorem tells us that, whatever that something is, it is almost surely not DNA, RNA, chromosomes, organisms, or resources. It tells us that there is good reason to believe that the things that we perceive, such as DNA and RNA, don't exist independent of our minds. The reason is that the structures of fitness payoffs, which shape what we perceive, differ from the structures of objective reality *with high probability*. Again, this is no support for solipsism: there is an objective reality. But that reality is utterly unlike our perceptions of objects in space and time.

Such a conclusion may seem absurd. Surely it's due to an error of logic. We just need to spot the error. Perhaps the error lurks in simplifying assumptions of evolutionary games. For instance, such games omit explicit mutations, assume an infinity of players, and stipulate that each player has an equal chance to compete with any other. These simplifications are generally false. Organisms in nature suffer mutations, have finite populations, and interact more with those close by.

Evolutionary games ignore these complexities and focus instead on the effects of natural selection. This is precisely the focus we need to test the claim that natural selection favors true perceptions. And the result, the FBT Theorem tells us, is clear: it doesn't.

An important process omitted by evolutionary games is neutral drift, in which a mutation that has no effect on fitness spreads by chance through a population. It might even drive other alleles to extinction. Such a mutation can mitigate the effects of natural selection, so that a difference in fitness that is decisive in evolutionary games is not decisive in a finite population with mutations. If, for instance, *Fitness* has a selective advantage over *Truth* of 100 percent, then, in an evolutionary game with an infinite population, *Truth* always goes extinct when competing with *Fitness*. But in a game with one hundred *Truth* players, the chance is only one-half that *Truth* goes extinct if a mutation introduces a *Fitness* player. This is a big difference.

But it's no boon for the claim that natural selection favors *Truth*. That claim is false, whether populations are finite or infinite, and whether mutations are explicit or not. A finite population can slow natural selection's annihilation of *Truth*—as blasting a bridge may slow an enemy tank—but cannot make it friendly.

If we wish to model different likelihoods of interactions between players, then evolutionary games must be played on graphs.[23] This theory is difficult and in its infancy. We know that networks of connections between players can amplify and dilute the pressures of natural selection in complex ways. There is much to be studied in this relatively new field. But so far, there is no support for the claim that natural selection favors *Truth*. The structure of a network may aid or retard the pressures of selection, but these pressures remain hostile to *Truth*.

Justin Mark, while a graduate student in my lab, used genetic algorithms, with explicit mutations, to study the coevolution of perception and action in finite populations.[24] He created an artificial world in which a player could forage for resources, and score fitness points. It could walk, look for resources, eat resources, and bump into walls that bounded the world. A suite of genes determined its actions and perceptions. The first generation of players had genes chosen at random, so that their actions and perceptions were haphazard, even comically stupid. Some would repeatedly hit a wall, or stay in one place, or repeatedly try to eat nothing. Each was so witless that, by the end of its foraging run, it had scored few points. But some were less daft than others. These were "bred" and their genes mutated to form a new generation. This process was repeated for hundreds of generations. By the last generation, all players foraged with efficiency and apparent intelligence. The question was: Did they evolve to see the truth?

The answer was no. Even when perception and action had coevolved for hundreds of generations, *Truth* did not appear. Players in the last generation saw the fitness of resources, but not their true quantities. Only on the off chance that fitness points track world structures could *Truth* appear.

These simulations do not constitute a proof. But they suggest that the extinction of *Truth* in evolutionary games cannot be pinned on faulty assumptions. Instead, *Truth* goes extinct because it hunts reality rather than fitness, like a chess player who hunts rooks rather than the king.

What other mistake may account for the conclusion that *Truth* goes extinct? Perhaps a notion of veridical perception that is too strong?

Consider three notions of veridical perception.[25] The strongest is "omniscient realism"—we see all of reality as it is. Next is "naive realism"—we see some, but not all, of reality as it is. The weakest is "critical realism"—the structure of our perceptions preserves some of the structure of reality. If the FBT Theorem targeted omniscient or naive realism, then we could indeed dismiss its conclusion—no one (save lunatics and solipsists) claims omniscience, and few espouse naive realism. But the theorem targets critical realism, which is the weakest, and most widely accepted, notion of veridical observation in the science of perception and in science more broadly. The FBT Theorem does not torch a straw man.[26]

Perhaps the theorem has made a mistaken assumption about objective reality? It proves that seeing reality leads to extinction. But what reality? And how could the theorem know or postulate, a priori, what reality is? A mistake on this point would surely defang the theorem.

Indeed it would. For the theorem to be of value, it cannot require a specific model of objective reality, but instead must be true in general. For this reason, the FBT Theorem assumes only that reality, whatever it is, has a set of states. States of what, the theorem does not say. It assumes only that states, or subsets of states, can have probabilities. But it specifies no particular probabilities.

The FBT Theorem asserts that if reality outside the observer has any structure beyond probability, then natural selection will shape perception to ignore it. The theorem makes no assumptions about the states of reality beyond the claim that we can discuss their probabilities. This claim could be false. But if it is, then a science of reality is impossible, for there would be no way to relate probabilistic outcomes of experiments to probabilistic claims about reality. Perhaps a science of reality is not possible. I hope otherwise. But the FBT Theorem, for its part, simply assumes that such a science is possible.

Perhaps the FBT Theorem is irrelevant to human evolution? Perhaps what is required to understand human evolution is a complete artificial-intelligence simulation of humans, together with a simulation of their interactions with all other organisms and with the earth itself. Perhaps, without such a comprehensive simulation, we cannot possibly claim to know that we did not evolve to see reality as it is.

Admittedly, our interactions with the environment are complex—indeed so complex that our evolution is chaotic: an infinitesimal nudge to the world now can trigger a tectonic transformation later. But the FBT Theorem still applies to human evolution.

An analogy can help us see why. Consider the state lottery. Millions of tickets are purchased by thousands of people for hundreds of different reasons, using dozens of different tricks for picking a particular number—birthdays, anniversaries, messages in fortune cookies. Suppose we wish to predict how many people will win at the next drawing. Do we need a complete simulation of all this complexity to get an answer? Not at all. Indeed, it would be a distraction. What is needed instead are a few principles of probability that apply regardless of the myriad details.

The same is true of the FBT Theorem. It allows us to guess, based on principles of probability, how many creatures will evolve to see reality as it is. The key insight of the theorem is simple: the probability that fitness payoffs reflect any structure in the world plummets to zero as the complexity of the world and perception soars. Chaotic effects prevent precise prediction of the specific perceptual systems that will prevail. But the laws of probability dictate that *Truth* has less chance than your lottery ticket.

Does this mean that our perceptions lie to us? Not really. I wouldn't say that our senses lie, any more than the desktop of my computer lies when it portrays an email as a blue, rectangular icon. Our senses, like the desktop interface, are simply doing their job, which is not to reveal the truth, but to guide useful actions. The FBT Theorem reveals that as the senses grow more complex, they have less chance to disclose any truths about objective reality.

Perhaps the FBT Theorem only holds for fixed payoffs? If payoffs fluctuate rapidly then perhaps the best strategy is to see reality as it is?

I grant that payoffs, like weather, are mercurial. And for the same reason—both arise from complex interactions among a plethora of factors. But protean payoffs afford *Truth* no purchase. *Truth*, no less than *Fitness*, must track the volatile sequence of fitness payoffs. At each step in this sequence, the FBT Theorem reveals, *Truth* is less fit—a negative amortization that hastens its ruin.

Although the flux of payoffs is no help to *Truth*, it does suggest that *Fitness* will be shaped by natural selection to report *differences* in payoffs rather than absolute payoffs. We see evidence of this in research on perceptual adaptation. Put on rose-colored glasses and the world looks reddish, but not for long. Soon you see the normal gamut of colors. Stare at a waterfall for a minute, then look at the rocks nearby. They appear to move up while also, paradoxically, staying put. Enter a movie theater on a sunny afternoon, and everything looks black. But soon you see shades of gray. Stare at a happy face for a minute, then look at a face with a neutral expression. It now looks sad. Stare at a blurry image for a few seconds and the world looks sharper; stare at a sharpened image and the world looks blurry. It was thought that adaptation is simply an anomaly due to overexposure. But experiments by the cognitive scientist Michael Webster reveal that it is an essential feature of all levels of perceptual processing.[27] Change the perceptual environment, put on rose-colored glasses, and your senses quickly adapt to report relative payoffs in the new context; they efficiently encode information about fitness.

Or you can fix the environment and change payoffs. Brian Marion, while a graduate student in my lab, had observers play a game in which they earned points for discriminating colors. If they were offered more points for discriminating blues than reds, then within minutes they better discriminated blues.[28]

This makes sense if perception reports differences in payoffs. Where there's no difference in payoffs there's no payoff in seeing differences. Where there are differences in payoffs then there is payoff in adjusting in real time to see those differences—not ideally or perfectly, just a bit better than the competition. Adaptation to scenes and rewards are two aspects of one process—tracking fitness payoffs. The reason that adaptation is not a curious anomaly, but instead appears at all levels of perceptual processing, is that tracking fitness payoffs is not a curious anomaly—it is the whole game.

But this emphasis on natural selection and adaptation raises a different objection, one spelled out by the psychologist Rainer Mausfeld: "the actual role of natural selection in the evolution of complex biological systems is far from obvious. . . . Evolutionary biology has, in more recent years, accumulated pervasive evidence that suggests that the vast majority of evolutionary change has rather little to do with natural selection." Mausfeld worries that the arguments discussed here take natural selection "as an almost exclusive factor regulating evolutionary change."[29]

Natural selection does indeed act in coordination with many collaborators. There is, as we have discussed, genetic drift—the chance spreading of a neutral allele, which has no effect on fitness, throughout a population. This is more likely in smaller populations. Such drift, some claim, accounts for most of molecular evolution.[30] It is possible that today's neutral drift might, as niches change, become tomorrow's game changer.

Then there is physics. Gravity, for instance, impedes the stability of moving limbs and the circulation of blood—inducing the evolution of bilateral symmetry in most animals and hindering the evolution of necks longer than a giraffe's. Then there is chemistry. Of the ninety-two elements that occur in nature only six—carbon, hydrogen, nitrogen, oxygen, calcium, and phosphorus—compose 99 percent of the mass of organisms. There is linkage: alleles nearby on a chromosome tend to be inherited together during meiosis. There is pleiotropy: one gene can influence disparate aspects of the phenotype, sometimes with opposing effects on fitness.

There are, no doubt, other factors in evolutionary change. And, for all I know, Mausfeld may be right that the vast majority of evolutionary change has rather little to do with natural selection. But this is no problem for the argument here. The question is not how much evolutionary change is due to natural selection, but rather about the direction of natural selection itself. No one argues, for instance, that we see reality as it is because of the evolutionary process of genetic drift. Genetic drift can't do the job. Nor can physics, chemistry, linkage, or pleiotropy. When proponents of veridical perception use evolution to argue for their view, they argue that veridical perceptions are *fitter* perceptions—that seeing reality as it is endows a selective advantage. Whether or not natural selection is the major force in evolution, it is the force that proponents of veridical perception appeal to—the only one, it would seem, that they can appeal to—in support of their claim.

What the FBT Theorem reveals is that natural selection, however major or minor a force it may be, does not shape our perceptions to be veridical. This is bad news for veridical perception in the only place where some had hoped the news might be good.

Perhaps the FBT Theorem has made a different, and quite fundamental, blunder. Philosopher Jonathan Cohen puts it as follows: "perceptual states have content—intuitively, what they carry information about, tell us about, or say about, the world, and that can be evaluated for truth or falsity."[31] So, for instance, if I have a perceptual experience that I describe as seeing a red tomato a meter away, then the content of my experience, what it says about the world, might be that in fact there is a red tomato a meter away. Indeed, that is a standard claim, in many philosophical accounts, about the content of such an experience.

But the FBT Theorem does not specify what the content of perceptual experiences might be. It simply concludes that experiences, whatever their contents, are not veridical.

Cohen argues that this is a blunder because "you can't say whether something is veridical or not without first knowing what it is saying."[32] So, if I say "one plus one equals two," you can decide if that statement is true because you know what it is saying. But if I say "blah plus blah blah," then you can't know if that statement is true because it is meaningless. It has no content.

If Cohen is right, then the FBT Theorem has made a fundamental error at the very start. It does not tell us, up front, what the contents of perceptual experiences are—what our experiences say about the world. So the theorem cannot possibly tell us whether our perceptual experiences are veridical. The theorem was a fool's errand from the start.

Fortunately for the FBT Theorem, there is no problem here. Philosophers have told us why, in their study of formal logic. Suppose that I tell you that *p* is some particular claim and *q* is some particular claim, but I refuse to tell you what either claim is. Then suppose I make the further claim, "*p* is true or *q* is true." If I ask you whether this last claim is true, you would have to shrug; if I don't reveal the contents of *p* and *q*, then, as Cohen says, you can't answer the question. But suppose that I instead claim, "if either *p* is true or *q* is true then it follows that *p* is true." And now I ask you if this claim is true. You don't have to shrug your shoulders. You know that this claim is false, even though you don't know the contents of *p* or *q*.

This is the power of logic, and of mathematics more generally. It allows us to evaluate the truth or falsity of large classes of statements simply in virtue of their logical or formal structure. Mathematicians prove theorems about functions and other structures on sets, without ever answering the question "Sets of what?" They don't care. It doesn't matter. Whether it is a set of apples, oranges, quarks, or possible universes, the theorems still apply. No prior content needs to be specified for the elements of the sets.

In particular, the rich field of information theory, which underlies the internet and telecommunications, has powerful tools and theorems detailing how messages can be structured and communicated—without ever specifying the content of any message.[33] The variety of particular contents is endless, but they all conform to specific rules, allowing us to create a rigorous science—information theory—that applies to all messages of any content. This insight underlies the FBT Theorem, which uses the formal structure of universal Darwinism to tell us universal facts about any evolved perceptual systems, regardless of their particular contents.

The FBT Theorem needs no prior theory of perceptual content. But in a reversal of the logic proposed by Cohen, the theorem actually *constrains* admissible theories of perceptual content. In particular, according to the FBT Theorem, any theory of content that assumes perceptions are, in the normal case, veridical is almost surely false, because we evolved to detect and act on fitness, not to perceive the true structure of objective reality. This applies to our perceptions of the middle-sized objects around us. When I have an experience that I describe as a red tomato a meter away, the content of that experience is not that there is—in objective reality, even when no one looks—a red tomato a meter away. As it happens, then, the FBT Theorem rules out all theories of content currently proposed in the philosophy of perception.[34]

The FBT Theorem extends an insight of the evolutionary theorist Robert Trivers: "the conventional view that natural selection favors nervous systems which produce ever more accurate images of the world must be a very naïve view of mental evolution."[35] It is also, according to the FBT Theorem, a very naïve view of perceptual evolution.

Steven Pinker sums up the argument well: "We are organisms, not angels, and our minds are organs, not pipelines to the truth. Our minds evolved by natural selection to solve problems that were life-and-death matters to our ancestors, not to commune with correctness."[36]

When the universal acid of Darwin's dangerous idea is poured onto our perceptions, it dissolves the objectivity of physical objects, which we assumed exist and interact even when no one looks. Then this acid dissolves the objectivity of spacetime itself, the very framework within which Darwinian evolution has been assumed to take place. This requires us to devise a more fundamental framework—without space, time, and physical objects—for understanding reality. We will need to understand the dynamics of this new framework. When we project this dynamics back into the spacetime interface of *Homo sapiens*, we should get back Darwinian evolution. Darwin's idea forces us to think of Darwinian evolution itself as an imperfect hint, couched within the spacetime-and-objects language of our perceptions, about a deeper, and as yet unknown, dynamics. Darwin's idea is indeed dangerous.

CHAPTER FIVE

Illusory

The Bluff of a Desktop

> "This is your last chance. After this, there is no turning back. You take the blue pill—the story ends, you wake up in your bed and believe whatever you want to believe. You take the red pill—you stay in Wonderland and I show you how deep the rabbit-hole goes."
>
> —MORPHEUS, *THE MATRIX*

I own life insurance. I'm betting there is an objective reality that exists even if I don't. If there is an objective reality, and if my senses were shaped by natural selection, then the FBT Theorem says the chance that my perceptions are veridical—that they preserve some structure of objective reality—is less than my chance to win the lottery. This chance goes to zero as the world and my perceptions grow more complex—even if my perceptual systems are highly plastic and can change quickly as needed.

This theorem is counterintuitive. How can my perceptions be useful if they aren't true? Our intuitions need some help here.

A venerable tradition conscripts the latest technology—clocks, switchboards, computers—to be a metaphor of the human mind. In line with this tradition, I invite you to explore a new metaphor of perception: each perceptual system is a *user interface*, like the desktop of a laptop. This interface is shaped by natural selection; it can vary from species to species, and even from creature to creature within a species. I call this the *interface theory of perception* (ITP). That name is a bit rich for a mere metaphor, but I try in what follows to pay the promissory note.[1]

Let's begin by digging deeper into an example from the preface. Suppose you're crafting an email, and the icon for the file is blue, rectangular, and in the center of the desktop. Does this mean that the file itself is blue, rectangular, and in the center of your computer? Of course not. The color of the icon is not the true color of the file. The shape and location of the icon are not the true shape and location of the file. Indeed, the file has no color or shape; and the location of its bits in the computer is irrelevant to the placement of its icon on the desktop.

The blue icon does not deliberately misrepresent the true nature of the file. Representing that nature is not its aim. Its job, instead, is to *hide* that nature—to spare you tiresome details on transistors, voltages, magnetic fields, logic gates, binary codes, and gigabytes of software. If you had to inspect that complexity, and forge your email out of bits and bytes, you might opt instead for snail mail. You pay good money for an interface to hide all that complexity—all that truth, which would interfere with the task at hand. Complexity bites: the interface keeps its fangs at bay.

The language of the interface—pixels and icons—cannot describe the hardware and software it hides. A different language is needed for that: quantum physics, information theory, software languages. The interface helps you craft an email, edit a photo, like a tweet, or copy a file. It hands you the reins of the computer and hides how things actually get done. Ignorance of reality can aid command of reality. This claim, out of context, is counterintuitive. But for an interface it's obvious.

ITP claims that evolution shaped our senses to be a user interface, tailored to the needs of our species. Our interface hides objective reality and guides adaptive behavior in our niche. Spacetime is our desktop, and physical objects, such as spoons and stars, are icons of the interface of *Homo sapiens*.

Our perceptions of space, time, and objects were shaped by natural selection not to be veridical—not to reveal or reconstruct objective reality—but to let us live long enough to raise offspring.

Perception is not about truth, it's about having kids. Genes that fashion perceptions that help us raise kids are genes that may win the fitness game and elbow their way into the next generation. The FBT Theorem tells us that winning genes do not code for perceiving truth. ITP tells us that they code instead for an interface that hides the truth about objective reality and provides us with icons—physical objects with colors, textures, shapes, motions, and smells—that allow us to manipulate that unseen reality in just the ways we need to survive and reproduce. Physical objects in spacetime are simply our icons in our desktop.

To ask whether my perception of the moon is veridical—whether I see the true color, shape, and position of a moon that exists even when no one looks—is like asking whether the paintbrush icon in my graphics app reveals the true color, shape, and position of a paintbrush inside my computer. Our perceptions of the moon and other objects were not shaped to reveal objective reality, but to disclose the one thing that matters in evolution—fitness payoffs. Physical objects are satisficing displays of crucial information about payoffs that govern our survival and reproduction. They are data structures that we create and destroy.

The language of space and time, of physical objects with shapes, positions, momenta, spins, polarizations, colors, textures, and smells, is the right language to describe fitness payoffs. But it is fundamentally the wrong language to describe objective reality. We cannot properly describe the inner workings of a computer in the language of desktops and pixels; similarly, we cannot describe objective reality in the language of spacetime and physical objects.

"But," you might say, "ITP has made a silly and obvious mistake: if a rattlesnake is just an icon of your interface, then why don't you grab one? After you're gone, and ITP with you, we'll know that our perceptions indeed tell us the truth."

I won't grab a rattlesnake, for the same reason I won't carelessly drag a paintbrush icon across my artwork in a graphics app. Not because I take the icon literally—there is no paintbrush in my laptop. But I do take it seriously. If I drag it around I could ruin my artwork. And that is the point. Evolution has shaped our senses to keep us alive. We had better take them seriously. If you see a fire, don't step in; if you see a cliff, don't step off; if you see a rattlesnake, don't grab; if you see poison ivy, don't dine.

I must take my senses seriously. Must I therefore take them literally? No. Logic neither requires nor justifies this move.

But we're inclined to say yes, and thereby fall prey to the *Serious-Literal fallacy*. Our specious conflation of serious and literal tempts us to reify physical objects and snipe-hunt among our figments for progenitors of consciousness. I understand the allure. I, too, feel the impulse to reify middle-sized objects. But I give it no credence.

Consider the biohazard and ionizing-radiation warning signs. Each must be taken seriously: ignoring either sign could be a last, and painful, mistake. But no one takes them literally: the biohazard sign does not depict biohazards as they are in objective reality, nor does the ionizing-radiation sign accurately depict ionizing radiation. Similarly, a sonar operator on a submarine must take seriously a green, glowing dot that streaks toward the center of the display. But torpedoes are not green, glowing dots. Evolution has shaped our perceptions with symbols, like a streaking green dot or a biohazard triangle, that warn us and guide us without depicting the truth.

So yes, if I see a rattlesnake writhing my way, I must take it seriously. But it doesn't follow that there is something brown, sleek, and sharp of tooth when no one observes. Snakes are just icons of our interface that guide adaptive behaviors, such as fleeing.

Such examples fail to convince some skeptics. Michael Shermer, for instance, in his column for *Scientific American*, wrote, "But how did the icon come to look like a snake in the first place? Natural selection. And why did some nonpoisonous snakes evolve to mimic poisonous species? Because predators avoid real poisonous snakes. Mimicry works only if there is an objective reality to mimic."[2]

Not so. Mimicry works if there is an *icon* to mimic. Consider the bird-dropping spider, *Celaenia excavata*, of eastern and southern Australia. It evolved to resemble excretions of its avian predators. Natural selection shaped the spider so that its icon in an avian interface approximates icons of droppings within that same interface. Indeed, one implication of ITP is that competition between predator and prey can trigger an evolutionary arms race between interfaces and interface hacks (such as masquerading as a dropping). We see an analogous arms race in phishing attacks on the internet, in which the logo, typography, and boilerplate of a legitimate bank or corporation are mimicked in an attempt to trick an unsuspecting victim into disclosing confidential information. A phishing attack that mimics, say, the Nike swoosh, doesn't work because Nike itself is, in objective reality, a swoosh. The swoosh is just an icon for Nike, and mimicking it can abet successful phishing, just as in nature mimicking an icon can hoodwink the interface of a predator or prey.

ITP predicts another head scratcher: a spoon exists only when perceived. Ditto for quarks and stars.

Why? A spoon is an icon of an interface, not a truth that persists when no one observes. My spoon is my icon, describing potential payoffs and how to get them. I open my eyes and construct a spoon; that icon now exists, and I can use it to wrangle payoffs. I close my eyes. My spoon, for the moment, ceases to exist because I cease to construct it. *Something* continues to exist when I look away, but whatever it is, it's not a spoon, and not any object in spacetime. For spoons, quarks, and stars, ITP agrees with the eighteenth-century philosopher George Berkeley that *esse is percipi*—to be is to be perceived.[3]

Let us revisit the Necker cube from chapter 1 (Figure 6). When you view the line drawing in the middle, you sometimes see a cube with face A in front,

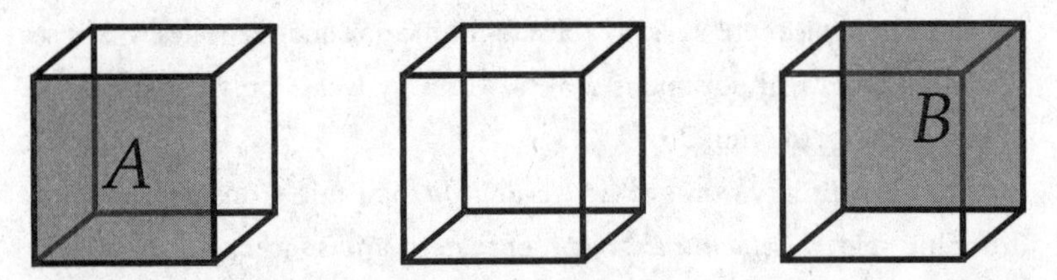

Fig. 6: The Necker cube. Which cube is there when you don't look? The cube with face A in front, or the cube with face B in front? © DONALD HOFFMAN

as shown on the left side of the figure. Call it Cube A. Other times you see a cube with face B in front, as shown on the right side of the figure. Call it Cube B. Now consider this question: Which cube is there in the middle when you don't look? Cube A or Cube B?

Well, it makes no sense to pick one over the other. Sometimes, when you look, you see cube A, sometimes cube B. The answer must be that, when you don't look, there is no cube—neither A nor B. Each time you look you see the cube you happen to construct at that time. When you look away, it goes away.

ITP says that the same is true for all objects in space and time. If you look and see a spoon, then there is a spoon. But as soon as you look away, the spoon ceases to exist. Something continues to exist, but it is not a spoon and is not in space and time. The spoon is a data structure that you create when you interact with that something. It is your description of fitness payoffs and how to get them.

This may seem preposterous. After all, if I put a spoon on the table then everyone in the room will agree that there is a spoon. Surely the only way to explain such consensus is to accept the obvious—that there is a real spoon, which everyone sees.

But there is another way to explain our consensus: we all construct our icons in similar ways. As members of one species, we share an interface (which varies a bit from person to person). Whatever reality might be, when we interact with it we all construct similar icons, because we all have similar needs, and similar methods for acquiring fitness payoffs. This is the reason we each see a cube in Figure 6—we each construct our own cube, but in much the same way as everybody else. The cube I see is distinct from the cube you see. I may see cube A at the same time you see cube B. There is no need to posit a real cube that everyone sees, and that exists when no one observes.

Indeed, there is no need to posit any physical object, or a spacetime, that exists when no one observes. Space and time themselves are simply the format of our interface, and physical objects are icons that we create on the fly as we attend to different options for collecting fitness payoffs. Objects are not preexisting entities that force themselves upon our senses. They are solutions to the problem of reaping more payoffs than the competition, from the multitude of payoffs on offer.

This is a new way of thinking about objects. We create them quickly, as needed, to solve fitness-gathering problems, and dispense with them just as quickly when they have, for the moment, served their purpose. They are not optimal solutions for grabbing payoffs, just satisficing solutions that let us nab a tad more than the competition.

Suppose I see a spoon, with some shape, color, texture, location, and orientation. In constructing this spoon, I solve a problem—I create a description of payoffs on offer and how to get them. I look away and the spoon disappears: my description of those payoffs is gone. I look back. I see a spoon again, because—no surprise—I've solved the same problem the same way. I can't help it. Natural selection has shaped me that way. I need fast solutions. I can't dally with novel techniques while rivals beat me to the punch. I have my go-to style for solving this problem, and in this context, I create a spoon every time. It's my habit.

I am inclined to reify my habit into an objective world. Why, I ask myself, do I keep seeing that spoon? Because, I tell myself, that spoon was there all along. Part of my logic is right. Something was there all along: my habit and an objective reality. But I'm wrong to assume that the objective reality is a spoon. I have made the mistake of reifying my habit into a preexisting spoon.

The Necker cube unmasks this kind of error. I look and see cube A. I look away and it disappears. I look back and, as it happens, I see cube B. It seems cube A wasn't actually there when I looked away. Something was there—my habitual way of creating descriptions of fitness payoffs. Normally it gives one description. In this case it offers two—which are similar, yet different enough that they could not be one preexisting object.

In like manner, I reify rocks, stars, and other icons in my interface, and pronounce them preexisting physical objects. I then reify the very format of my interface and fancy it to be a preexisting spacetime. This claim of ITP seems to agree with the philosophy of Immanuel Kant.[4] Exegesis of Kant is notoriously controversial, but one interpretation has him claim that rocks and stars are not mind-independent. They exist entirely in our perceptions.

Some philosophers find Kant's claim troubling. Barry Stroud, for instance, says, "What we thought was an independent world would turn out on this view not to be fully independent after all. It is difficult, to say the least, to understand a way in which that could be true."[5] To understand a way in which that could be true, we simply need to understand evolution by natural selection. According to the FBT Theorem, if selection shapes perceptions, then perceptions guide useful behaviors rather than report objective truths about an independent world. *Something* exists independent of us, but that something doesn't match our perceptions. This feels difficult to understand because of our penchant to reify our interface.

Kant also claims, as the philosopher Peter Strawson puts it, that "reality is supersensible and that we can have no knowledge of it."[6] On this point, ITP and Kant differ. ITP permits a science of objective reality. Kant, at least in some exegeses, does not. For scientists, this difference is fundamental. ITP asserts that one theory of objective reality—that it consists of physical objects in spacetime—is false. But ITP allows that the standard interplay of scientific theories and experiments could lead to a theory that is true. A first step is to recognize that our perceptions are an interface specific to our species, not a reconstruction of reality.

The biologist Jakob von Uexküll, in 1934, recognized that the perceptions of each species constitute a unique interface—an *umwelt*, as he puts it in the original German.[7] This accords with, and anticipates, ITP. But von Uexküll rejected the idea that each *umwelt* is shaped by natural selection, and proposed instead that its evolution is orchestrated according to a master plan. Here ITP and von Uexküll disagree. But they agree that rocks, trees, and other physical objects are icons of interfaces, not constituents of objective reality.

"But," you might say, "the claim that objects are icons creates a legal snafu. Suppose Mike drives a Maserati, and I'm jealous. I don't have that kind of money and probably never will. What to do? Suddenly I have the solution. Hoffman assures me that the Maserati is an icon I construct. That is, it's *my* icon! Well, what's mine is mine. I'll just take *my* icon for a joyride. In fact, I'll keep it. And no money down! After all, why should I pay for an icon that I construct? But alas, in fact there's just one Maserati here, one real public object that Mike and I see, and that exists even when no one looks. Mike paid for it, and I didn't, so I don't get to steal it. Too bad for ITP. Wish it were true. But ITP will land you in jail."

ITP does assert that the Maserati I see is just an icon I construct; there is no public Maserati. But ITP doesn't deny that there is an objective reality. It only denies that our perceptions describe that reality, whatever it is. Suppose an artist creates a digital masterpiece. From a remote location, I hack her computer and find her digital treasure. It appears as an icon on my desktop. *My* desktop, and *my* icon. So, since that icon is my icon, I reason that I can copy it and sell it. Clearly my reasoning is wrong. If I land in jail, I have myself to blame. Just because my icon is distinct from yours, and neither describes reality, it does not follow that I may do whatever I wish with my icon.

But if icons don't describe reality, are they real? What is real?

It's helpful to distinguish two different senses of real: existing, and existing even when unperceived.

If you claim that a Maserati is real, you probably mean that it exists even when no one looks. When Francis Crick wrote that the sun and neurons existed before anyone perceived them, he assumed that neurons are real in this sense. You need this assumption if you claim that neurons cause, or give rise to, our perceptual experiences. This assumption is denied by ITP and contradicted by the FBT Theorem.

If, however, I assert that I have a real headache, I claim only that my headache exists, not that it would exist even if unperceived. A headache that I don't perceive is no headache at all. I wouldn't mind that kind of "headache," of course. But if you tell me that my migraine is not real because it doesn't exist unperceived, I'm liable to become quite cross with you, and for good reason. My experiences are surely real to me, even if they don't exist unperceived.

Often the context will reveal what sense of "real" is at play. But to remove all doubt, it helps to say "objective" when discussing reality in the sense of existing unperceived. ITP asserts that neurons are not part of objective reality. They are, however, real subjective experiences—of a neuroscientist, for instance, peering at a brain through a microscope.

"But," you might say, "if the Maserati I see is not objective, why can I touch it when my eyes are closed? Surely that proves the Maserati is objective."

It proves nothing. It suggests, but does not prove, that there is *something* objective. But that something could be wildly different from anything you perceive. When you open your eyes, you interact with that unknown something and create a visual icon of a Maserati. When you close your eyes and reach out your hand, you create a tactile icon.

The same is true for all the other senses. If you close your eyes you may still hear the roar of an engine or smell the stench of exhaust. But these are your icons, and neither entails that the Maserati you perceive is part of objective reality.

"But if the Maserati I see is not objective, then why can my friend see it when my eyes are closed?"

There is an objective reality. You and your friend interact with it, whatever it might be, and each of you, in consequence, creates your own Maserati

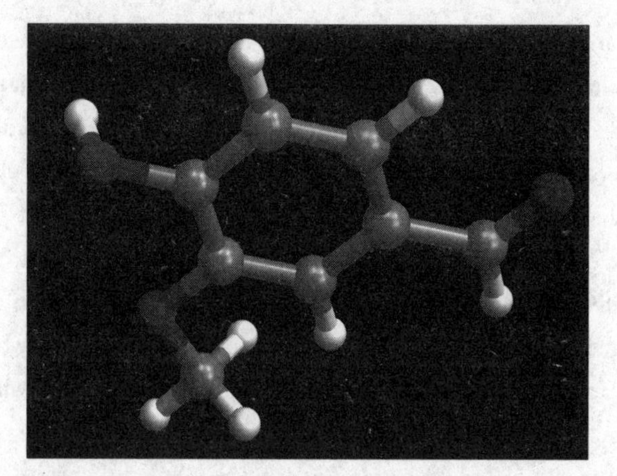

Fig. 7: A molecule with a special taste. © DONALD HOFFMAN

icon. It's not a problem for your friend to construct a Maserati icon when your eyes are closed, just as it's not a problem for her to construct cube A (or cube B) when your eyes are closed.

A red Maserati looks so shiny, artistic, aerodynamic, so *real*. But the FBT Theorem tells us that it's just a sensory experience—an icon—that is not objective and depicts nothing objective. Our intuitions rebel: our natural impulse is to reify Maseratis and other middle-sized objects. It's hard for us to let go of them. Fortunately, we find it much easier to let go of tastes. We happen to be less inclined to reify them. Let's see why, and perhaps this will help us resist the urge to reify middle-sized objects.

Consider the molecule depicted in Figure 7 and assume, for the sake of argument, that molecules are part of objective reality. The white spheres depict hydrogen atoms, the light gray spheres depict carbon, and the dark spheres depict oxygen. What sensory icon should you construct when you perceive this molecule? What taste experience accurately describes it?

These are not easy questions. Here are some clues. This is a phenolic aldehyde, an organic compound of molecular formula $C_8H_8O_3$, with functional groups aldehyde, hydroxyl, and ether.

So then, what taste truly describes this molecule? What taste most accurately depicts its true reality?

This molecule is vanillin. We perceive it as the delicious taste of vanilla. Who could have guessed? So far as I can tell, the taste of vanilla in no way describes that molecule. Indeed, no taste describes any molecule. Tastes are mere conventions. Yet tastes usefully inform our choices of what to eat, choices that could mean life or death.

If we had to check each atom before we chose what to eat, we would starve before vetting our dinner. The taste of vanilla, like tastes of all kinds, is a shortcut—an icon that guides our choice of cuisine. To ask whether the taste of vanilla describes $C_8H_8O_3$ is just as misguided as asking whether the letters CAT describe the furry pet, or the Maserati I see describes an objective reality.

In Plato's famous allegory of the cave, prisoners in the cave see flickering shadows cast by objects, but not the objects themselves.[8] This is a step in the direction of ITP, but it does not go far enough. A shadow vaguely resembles the object that casts it—the shadows of mice and men differ predictably in size and shape. The icons posited by ITP need resemble nothing of objective reality.

The shortcut of taste incurs a big risk—food poisoning. The solution hit on by evolution is to learn, in just one trial, to avoid a taste that is followed within hours by nausea. Your favorite food can, in one ill-fated day, become for years a trigger of disgust; the payoff you predict from its taste just went south.

The examples of vanillin and Maseratis are, of course, just examples. They prove nothing about perception and reality. That's the job of the FBT Theorem. But they may free us from our erroneous intuition that we see objective reality, and from our false belief that the moon is there when no one looks.

Some of my examples seem to backfire. Take the male beetles that conflate stubbies and female beauties. I trotted them out to show that evolution endows us with facile tricks and hacks that make us fit but hide the truth.

"But," you might retort, "they show the reverse. Why, according to Hoffman, is the beetle befuddled? Because, he claims, it can't see the truth. And how does he know that? Because he thinks *he* knows the truth—that the beetle really humps a bottle, not another beetle. So, hidden in his argument against seeing reality is the assumption that *he* sees reality, that he can tell a real beetle from a feigning bottle. Why else would he poke fun at the bungling beetle?"

This riposte seems compelling, but it fails. Suppose I watch a newbie playing *Grand Theft Auto*. He speeds a red Ferrari through the twisting curves of a mountain highway, oblivious to the ominous approach of a black helicopter. I shout a warning, but too late—his ride gets shredded by the blades of the chopper. I saw the folly of the newbie but not the "truth"—the transistors and voltages humming behind the glitz of the game. All I saw were icons, but I better understood what they meant. (The scare quotes on "truth" mean "truth for the sake of this example." Transistors and software are not objectively real.)

It's the same for the folly of beetles. I see icons of beetles and bottles, not objective truths. But my icons reveal a fact about fitness that the beetle's icons do not—humping bottles won't make baby beetles. Because my icons inform me of fitness, not truth, my critique of unfit beetle bumbling can be apt and yet presume no god's-eye view.

If icons are never true, are perceptions always illusions? The textbook account of illusions goes like this: "veridical perception of the environment often requires heuristic processes based on assumptions that are usually, but not always, true. When they are true, all is well, and we see more or less what is actually there. When these assumptions are false, however, we perceive a situation that differs systematically from reality: that is, an illusion."[9]

If our perceptions were normally veridical, then we could indeed define an illusion, such as the Necker cube, as a rare departure from truth. But ITP says that no perception is veridical, so it cannot define illusions this way. ITP does not, however, dismiss the notion of illusion: a Necker cube and a sugar cube are icons, but the two icons differ in some crucial way that must be understood. ITP needs a new account of illusions. And it has one, courtesy of evolution: *an illusion is a perception that fails to guide adaptive behavior.*

It's that simple. Evolution shapes our perceptions to guide adaptive behavior, not to see truth. So illusions are failures to guide adaptive behavior, not failures to see truth.

Let's take this theory for a spin. Why does ITP say that a beetle wooing a bottle suffers an illusion? Not because the poor beetle fails to see the truth. But because its perceptions prompt unfit actions: mating with bottles produces no beetles. Were it not for kind Australians who altered their stubbies, the beetles would have gone extinct.

Why, according to ITP, is the Necker cube an illusion? Because we cannot grasp in hand the shape we see. We can, by contrast, grasp a cube of sugar. One icon guides adaptive behavior and one does not. We are not, as it happens, deceived by the Necker cube. We know it is flat because its pictorial cues to depth are overruled by other visual cues, such as stereovision, that militate against any depth. This is to be expected. Our senses describe fitness payoffs and how to corral them. Getting this description right can mean life or death. So evolution equips us with multiple estimates. If they conflict, some estimates are given less credence or even ignored. There is safety in redundancy.

ITP's account of illusions obviates a nasty problem of the standard account. Consider the taste experiences of coprophagic animals—such as pigs, rodents, and rabbits. We can only hope that when they feast on feces their experiences differ markedly from our own. That they must differ is a clear prediction of ITP—tastes report fitness payoffs, not objective truths, with scrumptious tastes signaling better payoffs. The payoffs of feces, and thus their tastes, differ crucially between us and coprophages.

But this raises a baffling problem for the standard account, which claims that illusions are nonveridical perceptions: Whose perceptions are nonveridical—ours or those of coprophages? Are we right that feces truly have a loathsome taste? If so, do pigs, rabbits, and billions of flies suffer a taste illusion? Or are they right that feces truly are delicious? If so, is our disgusting experience a taste illusion?

Faced with such dilemmas, philosophers and psychologists sometimes answer that a perception is veridical if it is experienced by a standard observer under standard viewing conditions. A man who is red-green colorblind, for instance, when viewing grass under standard lighting, sees a color not seen by someone with normal color vision. So his colorblind perception is not veridical. It is tricky to specify standard observers and conditions in a principled way, and theorists twist themselves into pretzels trying. But here the gambit just won't work. To declare that humans are the standard is parochial. To defer instead to pigs and rabbits is to admit that feces in fact taste great. Neither choice is palatable. Feces pose a reductio ad absurdum of the theory that our perceptions are normally veridical, and that illusions are nonveridical perceptions.

The red berry of *Richadella dulcifica*, sometimes called the miracle berry, contains the glycoprotein molecule miraculin. If you eat this berry, then lemons and other sour foods taste sweet. The molecules of citric acid and malic acid in a lemon normally trigger a sour taste. But in the presence of miraculin, they trigger a sweet taste.

Which taste is illusory? The veridical-perception theory says it's the taste that's not veridical, that's not objectively true. So, what is the veridical taste of a molecule of citric acid? If we say it is sour, what is the ground for this claim? What principle requires a particular molecule to truly be a particular taste? The burden is on the veridical theorist to provide a scientific justification. None has been offered. Any claim of veridicality for any taste is, for now, thoroughly implausible.

ITP says that a taste is illusory if it prompts behaviors that are unadaptive. If, for instance, you've hunted gazelles all day and your blood sugar is low, you normally prefer foods that taste sweet, such as honey or an orange, and you're less inclined toward foods that taste sour, such as lemons. A lemon offers, gram for gram, half the calories of a sweet orange and one-tenth the calories of honey. In normal circumstances, a sweet taste guides adaptive eating that restores your blood sugar. But suppose you ate a miracle berry while hunting, so that a lemon tastes sweet. The sweet taste of the lemon now guides you to a poorer source of calories. It is less adaptive and thus illusory.

There is, it may seem, a more fundamental problem with ITP. It appeals to the FBT Theorem, which uses math and logic to prove that there's little chance we evolved to see objective reality. But what about our perceptions of math and logic? Doesn't the theorem assume math and logic, and then prove there's almost no chance that our perceptions of math and logic are true? If so, isn't it a proof that there are no reliable proofs—a reductio ad absurdum of the whole approach?

Fortunately, the FBT Theorem proves no such thing. It applies only to our perceptions of states of the world. Other cognitive capacities, such as our abilities with math and logic, must be studied on their own to see how they may be shaped by natural selection. It is too simplistic, and false, to argue that natural selection makes all of our cognitive faculties unreliable. This illogic is sometimes floated to support religious views believed to be incompatible with Darwinian evolution.[10] But it wields too broad a brush.

There can be selection pressures for modest facility with mathematics. The coin of the evolutionary realm is fitness, and counting that coin can be adaptive. Taking two bites from an apple provides roughly twice the fitness payoff as taking one. Because mathematics can aid reasoning about payoffs, selection is not uniformly against developing these talents. This is, of course, no argument that mathematics is an objective reality or that there are selection pressures for mathematical genius. It may be that such genius is a genetic fluke. Or perhaps sexual selection, in which the desires and choices of one sex shape the evolution of the other, can fan the flickers of basic mathematical skill into the flames of mathematical genius—a fascinating topic for research.

There can be selection pressures for modest facility with logic. For instance, social exchanges involve a simple logic of the form, "If I do this for you, then you must in return do that for me." Someone who cannot detect cheating in social exchanges is more likely to be fleeced, and thus less fit, than one who can detect cheating. So there are selection pressures for elementary ability with the if-then logic of these exchanges. Leda Cosmides and John Tooby have found that in most humans this ability with logic is less robust outside the context of social exchanges, where presumably it first evolved.[11] Similarly, the psychologists Hugo Mercier and Dan Sperber have found that our logical reasoning works best when we argue with others.[12] But once the basic ability is there, selection and mutation can take it to new places, even to the genius of a Kurt Gödel.

So, although ITP claims, and the FBT Theorem proves, that our perceptions of objects in spacetime do not reflect reality as it is, neither ITP nor the FBT Theorem preclude some skills with math and logic. Do they say anything about our higher conceptual skills? Do they entail that our concepts are likely to be the wrong concepts to understand reality as it is? Again, they do not. It remains an open question whether our species enjoys the concepts needed to understand objective reality. In chapter ten we consider a theory of reality that has the virtue that it allows, but does not require, that we possess the necessary concepts.

"But," one may wonder, "if I don't see reality as it is, then why does my camera see what I see? I drive to Yosemite Valley and head up to Tunnel View where I'm surrounded by scores of camera-toting tourists. I take the classic photo—El Capitan, Bridal Veil Falls, Half Dome—a breathtaking sculpture roughed out by a Sherwin glacier more than a million years ago and then chiseled to perfection by Tahoe, Tenaya, and Tioga glaciations. My photo matches what I see firsthand. It also matches what millions of others have seen and photographed. Surely this agreement can mean just one thing—we all see one ancient reality, and we see it as it really is. The camera doesn't lie."

This contention is psychologically compelling but logically unsound. Students in the life sciences can conduct experiments in virtual-reality labs, such as *Labster*, which offer a variety of virtual tools, such as microscopes, sequencers, and cameras. A student can grab a camera—an icon in the virtual lab—and snap a shot, confident that the camera sees what they see. But student and camera see nothing but icons. They agree, but neither sees objective reality.

Another concern lurks here, one raised by Michael Shermer in *Scientific American*. "Finally, why present this problem as an either-or choice between fitness and truth? Adaptations depend in large part on a relatively accurate model of reality. The fact that science progresses toward, say, eradicating diseases and landing spacecraft on Mars must mean that our perceptions of reality are growing ever closer to the truth, even if it is with a small 't.'"[13]

The either-or choice between fitness and truth is, as we have discussed, not a whim of ITP, but an essential feature of evolutionary theory—fitness payoffs are distinct from objective reality and can, for a given element of reality, vary wildly from creature to creature and time to time. To track fitness is simply not, in general, to track truth.[14]

But as Shermer notes, science makes progress. It learns to cure disease, explore the stars, and land on Mars. Cell phones and driverless cars would look like magic to a visitor from the nineteenth century. Technology grows ever more adept at controlling our world. Doesn't this mean that "our perceptions of reality are growing ever closer to the truth"?

Not at all. Players of *Minecraft* grow ever more adept at dealing with its worlds. But they do so by mastering an interface, not by growing ever closer to the truth. To a neophyte, an expert at *Minecraft* looks like a magician, but that expert may know nothing of the complex machinery that lurks behind the icons.

Scientific theories, couched in the language of objects in spacetime, are theories still bound to the interface. They can't properly describe reality any more than a theory couched in the language of pixels and icons can properly describe a computer. Some physicists, as we shall see, recognize this and have concluded that "spacetime is doomed" along with its objects.

Our prowess with diseases, spacecrafts, and cameras is impressive. But prowess is just prowess, not truth. We have become better masters of our interface. But as long as our theories are stuck within spacetime, we cannot master what lurks behind.

"But wait," you might say, "there's nothing new here. Ever since 1911, when Ernest Rutherford discovered that the atom is mostly empty space, with just a tiny nucleus at its center, physicists have told us that reality is quite different from what we see. That hammer may look solid but, if you look closely enough, you'll find that it too is mostly empty space, with electrons and other particles whizzing about at incredible speeds."

Indeed. But this claim of physicists is not as radical as the claim of ITP. Their claim is more like saying, "I know that the icons on my desktop are not the true reality. But if I pull out my trusty magnifying glass and look really closely at the desktop, I see tiny pixels. And those tiny pixels, not the big icons, are the true nature of reality."

Well, not really. Those pixels are still on the desktop, still in the interface. They may not be visible without a magnifying glass, but they're part of the interface nonetheless. Similarly, atoms and subatomic particles are not visible without special equipment, but they're still in space and time, and so they are still in the interface.

Physics reveals that we often fail to notice what is too fast or slow, too big or small, or simply outside the band of electromagnetic waves that we can see. ITP is saying something much deeper. It says that even though we can, with the help of technology, observe all these new things, we are no closer to seeing reality as it is. We are just exploring more of our interface, more of what happens within the confines of space and time.

These claims of ITP are indeed radical, and in making them ITP reaches beyond its origins in evolution and neuroscience, and trespasses into the turf of physics. Perhaps it has overreached. Perhaps the counterintuitive claims of ITP are readily rebuffed by theory and experiment in modern physics.

Let's see.

CHAPTER SIX

Gravity

Spacetime Is Doomed

"Einstein never ceased to ponder the meaning of the quantum theory. . . . We often discussed his notions on objective reality. I recall that during one walk Einstein suddenly stopped, turned to me and asked whether I really believed that the moon exists only when I look at it."

—ABRAHAM PAIS, *EINSTEIN AND THE QUANTUM THEORY*

"It means buckle your seatbelt, Dorothy, 'cause Kansas is going bye-bye."

—CIPHER, *THE MATRIX*

If our senses were shaped by natural selection, then the FBT Theorem tells us we don't see reality as it is. ITP tells us that our perceptions constitute an interface, specific to our species. It hides reality and helps us raise kids. Spacetime is the desktop of this interface and physical objects are among its icons.

ITP makes bold and testable predictions. It predicts that spoons and stars—all objects in space and time—do not exist when unperceived or unobserved. Something exists when I see a spoon, and that something, whatever it is, triggers my perceptual system to create a spoon and to endow it with a position, a shape, a motion, and other physical properties. But when I look away, I no longer create that spoon and it ceases to exist, along with its physical properties.

ITP predicts, for instance, that a photon, when unobserved, has no definite value of polarization. It predicts that an electron, when unobserved, has no definite value of spin, position, or momentum. An experiment that contradicted these predictions would disconfirm ITP.

The objects I see are my icons. The objects you see are your icons. When we compare notes, we find that our icons often agree—I see a cat and so do you; I see a fire and so do you. We often agree because we interact with the same reality, whatever it might be, and we deploy similar interfaces with similar icons. But ITP predicts that we can disagree. I may see fire, and cook my dinner, where you see none and your dinner stays cold; I may see a cat alive, where you see it dead.

ITP predicts that spacetime does not exist unperceived. My spacetime is the desktop of my interface. Your spacetime is your desktop. Spacetimes vary from observer to observer, and some properties of spacetime need not always agree across observers. Reality, whatever it might be, escapes the confines of spacetime.

These are, as I said, bold predictions. But are they really testable? Can they be ruled out by modern physics? I might boldly predict that the moon turns into swiss cheese when no one looks if I knew that my prediction could never be tested. To say that an electron has no spin when it is unobserved may sound bold, but how could this claim be tested? Can we perform an experiment, a careful observation, that tells us what happens when no one observes? If this sounds impossible to you, then, as I mentioned in chapter four, you're in good company, for it also seemed impossible to the brilliant physicist Wolfgang Pauli. Einstein worried whether quantum theory entails that "the moon only exists when I look at it." Pauli replied, "One should no more rack one's brain about the problem of whether something one cannot know anything about exists all the same, than about the ancient question of how many angels are able to sit on the point of a needle. But it seems to me that Einstein's questions are ultimately always of this kind."[1]

Einstein believed that spacetime and objects exist and have definite properties whether or not they are observed. More precisely, he believed in *local realism. Realism* is the claim that physical objects have definite values of physical properties—such as position, momentum, spin, charge, and polarization—even when unobserved. *Locality* is the claim that physical objects cannot influence each other faster than the speed of light. Local realism asserts that both realism and locality are true. Einstein insisted, as he wrote in a letter to the physicist Max Born, that physics should adhere to "the requirement for the independent existence of the physical reality present in different parts of space."[2] Einstein believed that quantum theory, which violates this requirement, must be an incomplete theory of reality. He noted, in his letter to Born, that "I still cannot find any fact anywhere which would make it appear likely that that requirement will have to be abandoned."[3]

That was true when Einstein wrote it in 1948. But in 1964, the physicist John Bell discovered a fact that would have stunned Einstein: there are experiments for which quantum theory predicts outcomes that contradict local realism.[4] Whether or not quantum theory is, as Einstein claimed, incomplete, it is incompatible with local realism. Bell's experiments have now been performed in multiple variations, and the predictions of quantum theory have been confirmed each time. We now have excellent evidence that local realism is empirically false, even if quantum theory is false or incomplete. This means that realism is false, or locality is false, or both are false. There is no happy choice here for Einstein, or for our normal intuitions.

One experimental test of local realism, inspired by Bell and conducted at the Delft University of Technology in the Netherlands, measured the spins of entangled electrons.[5] Electron spin is strange. Frisbees, tops, and ice skaters can spin slowly, quickly, or anywhere in between. Not an electron. If you measure its spin along any axis, you find that there are just two possible answers—up or down. It's as though the electron can spin either clockwise or counterclockwise, but at only one speed.

Entanglement is also strange. Place two spinning tops side by side, and you can describe each top and its spin separately. But you can't do that for two entangled electrons. They have to be described as though they were one indivisible object, no matter how distant they are from each other. For instance, a physicist can entangle the spins of two electrons so that if the spin of one electron along some axis is up, then the spin of the other electron along that axis is always down. This holds no matter which axis you choose to measure. It also holds no matter how far apart the electrons are. They could be a billion light-years apart. Still, if you measure the spin of the electron near you, then you instantly know what you would find if you measured the spin of the other electron a billion light-years away. If realism is true, and if your measurement of spin here instantly affects the spin of an electron a billion light-years away, then this effect violates the claim of locality—that no influence can propagate faster than the speed of light.

In the Delft experiment, two electrons separated by 1,280 meters had their spins entangled.[6] It takes light just over four millionths of a second to travel this distance. The spins of the two electrons were measured along randomly chosen axes. Critically, the two spins were measured at the same time. This assured that one measurement could not affect the other by any local process—that is, by a process that propagates no faster than the speed of light. The Delft experiment, like all the others, confirmed the predictions of quantum theory and rejected local realism. The spin measurements of the two electrons were correlated in a way that Bell showed would be impossible if local realism were true. Either realism is false, and the electrons had no definite values of spin before they were measured, or locality is false, and the electrons influenced each other at speeds faster than light. Or realism and locality are both false.

Physicists are trying to discern which assumption is false, realism or locality. Experiments with entangled photons by Anton Zeilinger and his collaborators have ruled out a large class of theories that claim that realism is true and locality is false.[7] They conclude, "We believe that our results lend strong support to the view that any future extension of quantum theory that is in agreement with experiments must abandon certain features of realistic descriptions."[8] Although the jury is still out, defending realism has gotten harder, thanks to the experiments of Zeilinger.

ITP predicts that realism is false, and physics does not contradict this prediction. Instead, each test of local realism, in defiance of our intuitions, confirms the prediction of ITP. Experiments such as Zeilinger's are tightening the noose around the neck of realism.

So is another theorem that follows from quantum theory and makes no assumption about locality. It was proven by Bell, in 1966, and by Simon Kochen and Ernst Specker, in 1967, and is called the Kochen-Specker (KS) Theorem. It says that no property, such as position or spin, has a definite value that is independent of how it is measured.[9] The opposite claim, that a property can have a definite value that is independent of how it is measured, is called "noncontextual realism." The KS Theorem says that noncontextual realism is false.

But noncontextual realism is precisely what we espouse in saying the moon is there when no one looks. It's the realism that Francis Crick had in mind when he wrote that the sun and neurons exist when no one looks. It is this realism that is false—independent of any issues about locality.

The KS Theorem shatters another belief that Einstein had about reality. In 1935, in a famous paper with Boris Podolsky and Nathan Rosen, he claimed that "If, without in any way disturbing a system, we can predict with certainty (i.e., with probability equal to unity) the value of a physical quantity, then there exists an element of reality corresponding to that quantity."[10]

This claim may seem plausible. Suppose you can tell me with total confidence, before you make a measurement, that the spin of an electron along some axis will certainly be observed to be up—there's no chance, you assure me, that it will be down. And suppose you're right every time, for thousands of observations. Then I may conclude that your confidence is warranted, and your prediction is always right, because the electron really had that spin all along.

But I would be wrong. The physicists Adán Cabello, José M. Estebaranz, and Guillermo García-Alcaine constructed a clever case of the KS Theorem.

In their example, quantum theory predicts the measured value of a physical quantity with certainty, "with probability equal to unity." But they prove that the value cannot exist independent of the measurement.[11] This means I can be certain what value I'll find, and yet that value is not an element of objective reality. Certainty about what you'll see doesn't imply it already exists. Einstein, Podolsky, and Rosen were simply wrong to claim otherwise.

Most of us believe deeply in a physical reality, consisting of objects in spacetime that existed prior to life and observers; no observer is needed, we believe, to endow any object with a position, spin, or any other physical property. But as the implications of quantum theory are better understood and tested by experiments, this belief can survive only by clinging to possible gaps in the experiments, and those gaps are closing. An experiment at Fermilab, for instance, reveals that neutrinos—subatomic particles with almost no mass—have no value of the physical property of lepton flavor until they are observed.[12]

Some physicists conclude that quantum theory counsels a profoundly new view of the world. As the physicist Carlo Rovelli puts it, "My effort here is not to modify quantum mechanics to make it consistent with my view of the world, but to modify my view of the world to make it consistent with quantum mechanics."[13] The way that Rovelli updates his worldview is to reject "the notion of an absolute, or observer-independent, state of a system; equivalently, the notion of observer-independent values of physical quantities."[14] Rovelli abandons noncontextual realism.

He explains why: "If different observers give different accounts of the same sequence of events, then each quantum mechanical description has to be understood as relative to a particular observer. Thus, a quantum mechanical description of a certain system . . . cannot be taken as an 'absolute' (observer-independent) description of reality, but rather as a formalization, or codification, of properties of a system relative to a given observer. . . . In quantum mechanics, 'state' as well as 'value of a variable'—or 'outcome of a measurement'—are relational notions."[15]

The physicist Chris Fields discards noncontextual realism on different grounds. He shows that if no observer sees all of reality at once, and if observing takes energy, then noncontextual realism must be false.[16] The physicists Chris Fuchs, David Mermin, and Rüdiger Schack claim that quantum theory entails "that reality differs from one agent to another. This is not as strange as it may sound. What is real for an agent rests entirely on what that agent experiences, and different agents have different experiences."[17] They explain, "A measurement does not, as the term unfortunately suggests, reveal a pre-existing state of affairs. It is an action on the world by an agent that results in the creation of an outcome — a new experience for that agent. 'Intervention' might be a better term."[18]

On Fuchs's interpretation of quantum theory, known as Quantum Bayesianism (or QBism), quantum states describe not the objective world but the beliefs of agents about consequences of their actions. Different agents may entertain different beliefs. No quantum state is universally true. Each is personal. My quantum state describes, as Chris Fuchs puts it, " 'The consequences (for *me*) of *my* actions upon the physical system!' It's all 'I-I-me-me mine,' as the Beatles sang."[19]

This agrees with the interface theory of perception. My perceptions of spacetime and objects are an interface, shaped by natural selection not to reveal reality but to guide my actions in ways that enhance my fitness. *My* fitness. What benefits me may harm another. A bar of chocolate that boosts my health could kill my cat. Natural selection shapes perceptions in a personal fashion, to tell *me* the consequences for *me* of *my* actions upon the world. There is a world that exists even if I don't look: solipsism is false. But my perceptions, like observations in quantum theory, don't disclose that world. They counsel me—imperfectly, but well enough—how to act to be fit.

Quantum theory and evolutionary biology, so interpreted, together weave a remarkably consistent story. Quantum theory explains that measurements reveal no objective truths, just consequences for agents of their actions. Evolution tells us why—natural selection shapes the senses to reveal fitness consequences for agents of their actions. We are surprised that measurement and perception are so personal. We expected them to report objective and impersonal truths, albeit fallibly and in part. But when two pillars of science side with each other, and against our intuition, it's time to reconsider our intuition.

This confluence of physics and evolution has not been obvious. In 1987, William Bartley described a conference in which the physicist John Wheeler presented his take on quantum theory. Sir Karl Popper, a famous philosopher of science, "turned to him and quietly said: 'What you say is contradicted by biology.' It was a dramatic moment. . . . And then the biologists . . . broke into delighted applause. It was as if someone had finally said what they had been thinking."[20]

Bartley tells us what the biologists were thinking: "Sense perceptions or sensations are themselves only more or less accurate symbolic representations of external reality formed through the interaction between that external reality and organs of sense. One sees external reality, more or less accurately."[21] This belief is no surprise. Evolutionary biology, as we have discussed, assumes the objective reality of objects such as DNA and organisms. It is not obvious that the acid of universal Darwinism—in the form of the FBT Theorem—dissolves this extraneous assumption and reveals that "more or less accurate symbolic representations of external reality" are never more fit than representations that hide external reality and encode fitness payoffs.

What did Wheeler propose that vexed the biologists? Wheeler claimed that, "What we call 'reality,' consists of an elaborate papier-mâché construction of imagination and theory filled in between a few iron posts of observation."[22] We don't, according to Wheeler, passively observe a preexisting objective reality, we actively participate in constructing reality by our acts of observation. "Quantum mechanics evidences that there is no such thing as a mere 'observer (or register) of reality.' The observing equipment, the registering device, 'participates in the defining of reality.' In this sense the universe does not sit 'out there.' "[23]

Wheeler illustrated this with his delayed-choice experiment, a variation of the famous double-slit experiment first conducted by the physicists Clinton Davisson and Lester Germer in 1927.[24] Recall that in the double-slit experiment a photon gun shoots one photon at a time toward a photographic plate that records where each photon lands. But between the gun and the plate is a metal screen with two tiny slits in it—call them *A* and *B*—through which the photons can pass.

If just one slit is open, then the photons land, as expected, on a portion of the photographic plate just behind that slit. But if both slits are open, then the photons land, contrary to expectations, in a sequence of bands reminiscent of the interference patterns one gets when two water waves collide—with the remarkable consequence that some locations on the plate that get lots of photons when just one slit is open, will get fewer photons, or even none, when both slits are open. In this case it appears, at first glance, that each photon somehow went through both *A* and *B* at the same time. That is no problem for a wave. But a photon is a particle; and if we do this same experiment with electrons, which are also particles, we get the same interference pattern.

So how does a particle do this trick? Does it split itself in half? If we try to observe the slits closely, we always see a photon go through just one slit, never both. Moreover, if we observe which slit it goes through then the interference pattern disappears.

No one really knows what a photon or electron does when both slits are open. This is an unsolved mystery of quantum theory. It seems incorrect to say it goes through *A*, through *B*, through both, or through neither. Physicists just say that its path is a *superposition* of *A* and *B*. This just means we don't know what's happening, even though we can write down simple formulas, involving linear combinations called superpositions, that accurately model the results of experiments. And it's not just tiny particles, like photons and electrons, that do this magic with double-slits. In 2013, Sandra Eibenberger and her collaborators found the same magic feat performed by a large molecule—fondly called C284.H190.F320.N4.S12—consisting of 810 atoms, and weighing more than 10,000 protons or 18 million electrons. It is a tad smaller than a virus.[25] Quantum weirdness is not confined to the subatomic realm.

Wheeler's delayed-choice variation on this experiment is clever: wait until after the photon passes the metal screen, and only then decide what to measure—path *A*, path *B*, or a superposition. In his words, "Let us wait until the quantum has *already* gone through the screen before we—at our free choice—decide whether it *shall have* gone 'through both slits' or 'through one.' "[26] Wheeler's experiment has been performed with photons (and helium atoms!) and it works.[27] What we choose to measure *after* the photon has passed the screen determines what the photon did, or at least what we can say about what it did, *before* we measured. "In the delayed-choice experiment we, by a decision in the here and now, have an irretrievable influence on what we will want to say about the past—a strange inversion of the normal order of time."[28] The past depends on our choice in the present. No wonder that Popper and the biologists were nonplussed.

Wheeler later expanded his experiment to cosmic scales.[29] Instead of a photon gun, consider a distant quasar—a supermassive black hole that sucks material from a surrounding galaxy into its accretion disk and, in the process, emits an astronomical amount of light and radiation, perhaps one hundred times the entire output of our Milky Way galaxy. Suppose this quasar lies behind a massive galaxy. According to Einstein's theory of gravity, such a galaxy bends spacetime. His theory also predicts that if everything lines up just right, we can see two images of that quasar, because its light can travel two different paths through the bent spacetime—a cosmic optical illusion caused by an enormous gravitational lens. Figure 8 shows an example in a photograph taken by the Hubble Space Telescope of the Twin Quasar QSO 0957+561, almost 14 billion light-years from earth.

With this, we have the setup needed for a delayed-choice experiment on a cosmic scale. Using a telescope to capture photons from the Twin Quasar, we can choose to measure which path through the gravitational lens a photon takes—the upper or lower path in the Hubble image—or we can choose to measure a superposition. If we choose to measure its path and we discover, say, that it's on the upper path, then for almost 14 billion years that photon has been on that path because of a choice we made today. If we had chosen instead to measure a superposition, then that photon would have a different history for the last 14 billion years. Our choice today determines billions of

Fig. 8: Image of Twin Quasar QSO 0957+561 taken by the Hubble Space Telescope. Credit: ESA/NASA

years of history. Most of us can't bench-press a hundred kilos. But we can reach back billions of years and trillions of kilometers to rewrite the past—a Herculean feat.

This raises the stakes. Quantum theory smashed our intuitions about objects, by denying that they have definite values of physical properties that are independent of whether, or how, they are observed. Now it smashes space and time. As Wheeler put it, "No space. No time. Heaven did not hand down the word 'time'. Man invented it. . . . If there are problems with the concept of time, they are of our own creation . . . as Einstein put it 'Time and space are modes by which we think, and not conditions in which we live.' "[30]

Einstein showed that different observers, moving at different speeds, disagree in their measurements of time and distance. But they agree about the speed of light, and about intervals in spacetime—a union of space and time into a single entity in which space and time can trade off. This raised the hope that spacetime is an objective reality even if space and time, separately, are not. Wheeler, wielding his delayed-choice experiment as a weapon of commonsense destruction, leveled this hope. "What are we to say about that weld of space and time into spacetime which Einstein gave us in his 1915 and still standard classical geometrodynamics? . . . no account of existence can ever hope to rate as fundamental which does not translate all of continuum physics into the language of bits."[31] He argued that spacetime and its objects are not fundamental. Instead he proposed the doctrine of "It from bit": information, not matter, is fundamental; the "its" of matter arise from bits of information. Wheeler's jump from spacetime to bits of information is more than a bit jarring. Why should the two be related? And why should bits replace spacetime? Spacetime seems so real—indeed the very bedrock and framework of reality. Surely spacetime existed before there were bits, and surely bits exist inside spacetime, not vice versa?

But once again our intuitions are wrong. An example reveals how wrong. Suppose I work for a computer manufacturer, and I have to design the memory for their next supercomputer. I want to cram the most memory into the least volume. The competition is stiff, so I want to get it right. I learn through the grapevine that my top competitor plans to cram its memory into six equal spheres, as shown in Figure 9. I smile. They've made a silly mistake. Those six spheres pack neatly into a larger sphere with more volume—in fact, over twice the volume. That larger sphere should hold over twice the memory. The competition is wasting all that valuable space between its six spheres. I'll use it to cram in more memory. I proudly tell the marketing department to get the ads ready—our computer has twice the memory of the competitor's.

But I'm wrong. If I and my competitor cram as much memory as possible into our designs, mine ends up with less memory—about 3 percent less. Even though my big sphere has twice the volume of their six smaller spheres combined, even though it could contain all six smaller spheres inside it, still, it holds less memory. If this bothers you, then you understand the problem.

Jacob Bekenstein and Stephen Hawking showed that the amount of information you can cram into a region of space is proportional to the area of the surface surrounding that space.[32] That's right, the *area*, not the volume.

Fig. 9: Six spheres packed inside a larger sphere. The six smaller spheres can hold more information than the larger sphere that surrounds them. © DONALD HOFFMAN

They first discovered this rule for black holes, but then realized it holds for any region of spacetime, not just regions containing a black hole. This rule is called the "holographic principle."

Hawking figured out how many bits of information an area can contain. To understand his result, you must first know that spacetime, like the desktop of your computer, has pixels—the smallest patches of spacetime that are possible. Smaller than that, spacetime simply doesn't exist. Each pixel of spacetime has the same length, called the Planck length.[33] It's tiny—about as tiny compared to a proton as the United States is to the entire visible universe. Spacetime also has a smallest area, called the Planck area, which is the square of the Planck length. These are the tiniest pixels of spacetime area that are possible. And Hawking discovered it's the number of these pixels in a surface, not the number of voxels in the volume inside, that dictates how many bits it can hold.

We all have strong convictions about space and time. Mine were stunned by the holographic principle. But I soon realized that this result fits well with

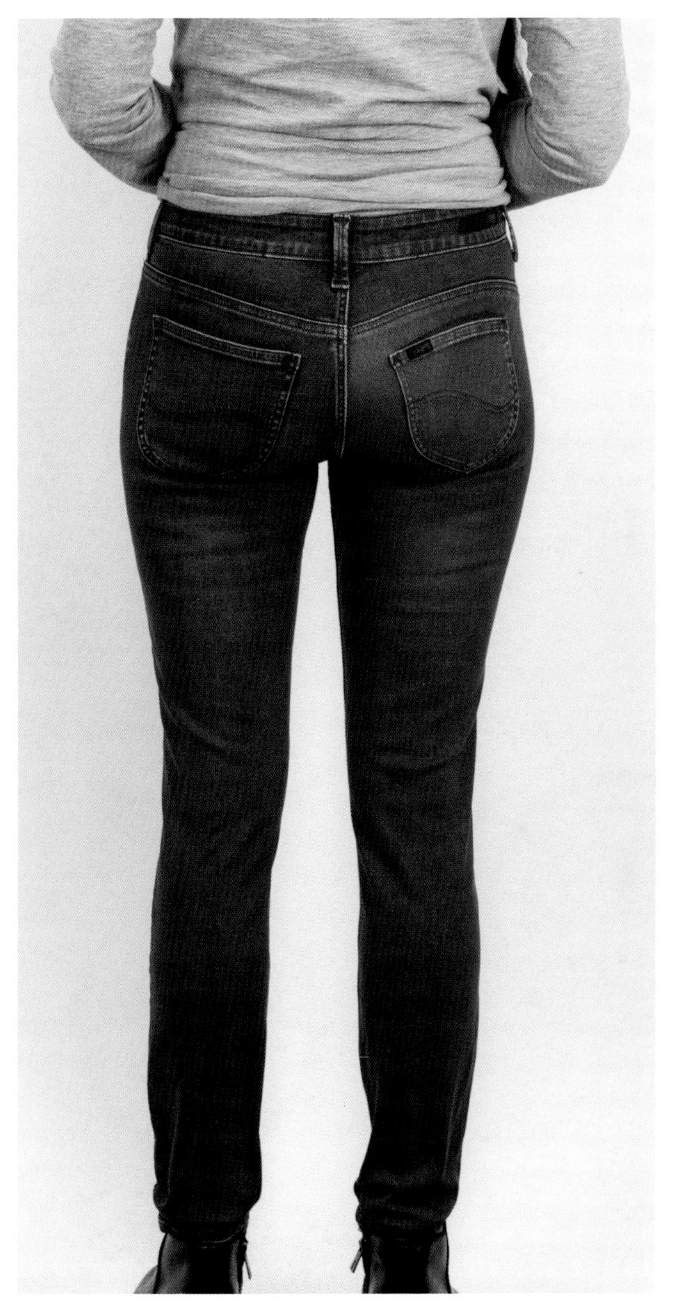

Fig. A: Enhancing the body with jeans. The left side looks flat. The right side looks firm and toned. The difference is due to careful use of visual cues for depth.

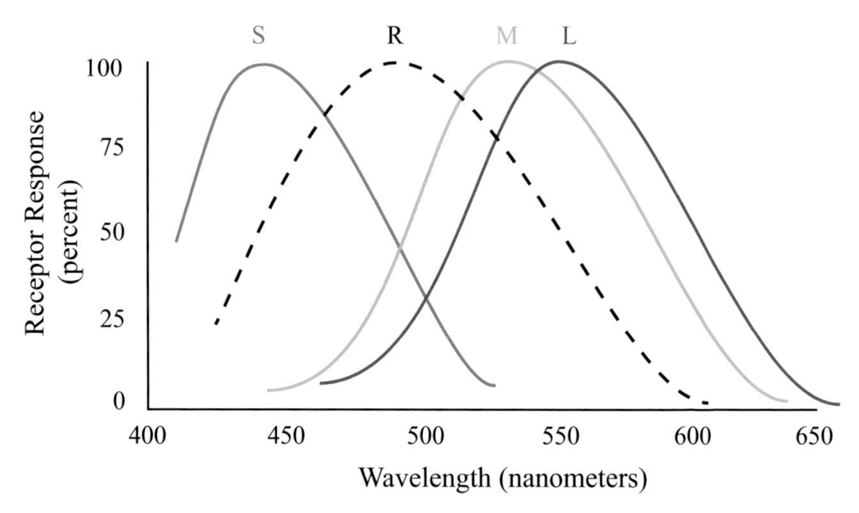

Fig. B: Sensitivity curves for the three types of cones in the retina of the eye (L, M, and S). The sensitivity of rods, which mediate vision in low light, is given by the "R" curve.

Fig. C: The Olympic rings illusion. The colors that fill each ring are illusory. The visual system creates them to correct an erasure error. © DONALD HOFFMAN

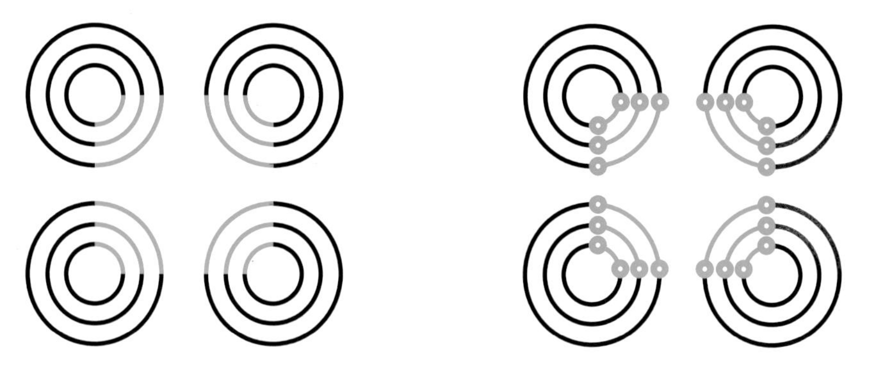

Fig. D: The neon square illusion. The glowing blue square is illusory. The visual system creates it to correct an erasure error. © DONALD HOFFMAN

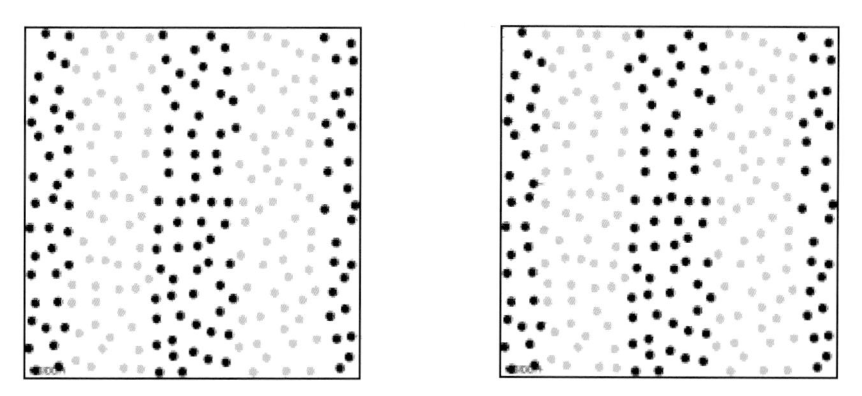

Fig. E: Two frames of dots from a movie. When the frames are displayed as a movie, the visual system creates blue bars that move, glow, and have sharp edges. © DONALD HOFFMAN

Fig. F: Joseph's hat illusion. The brown rectangle on the left side of the hat is printed in the same color ink as the yellow rectangle on the front of the hat. © DONALD HOFFMAN

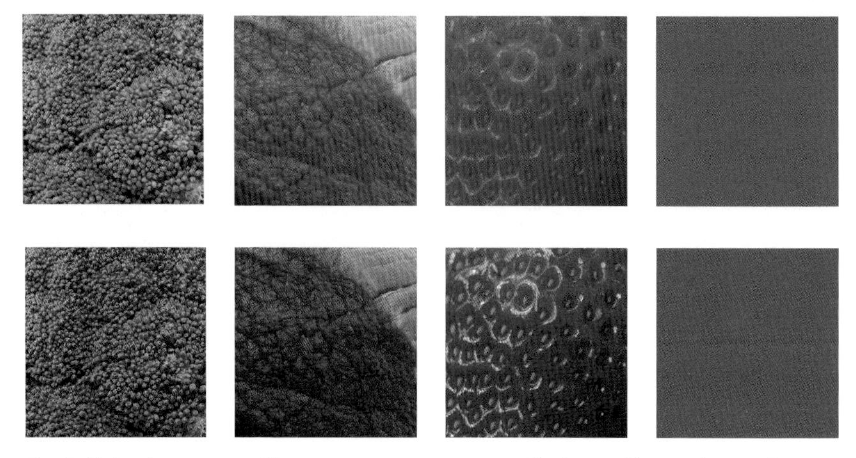

Fig. G: Eight chromatures. Chromatures are more versatile than uniform color patches at triggering specific emotions. © DONALD HOFFMAN

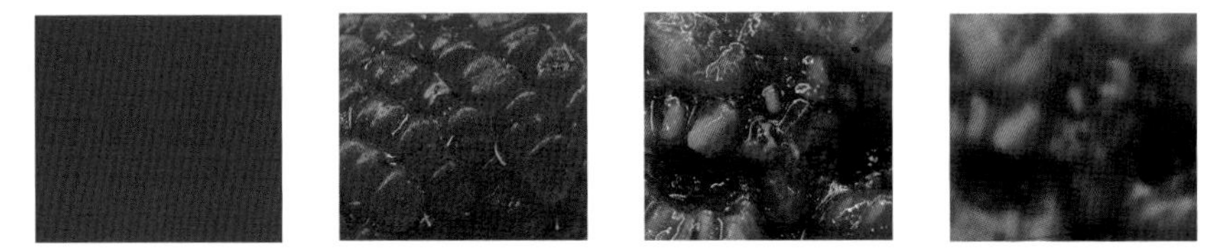

Fig. H: Four red chromatures. Red only triggers hunger if the texture is appropriate.

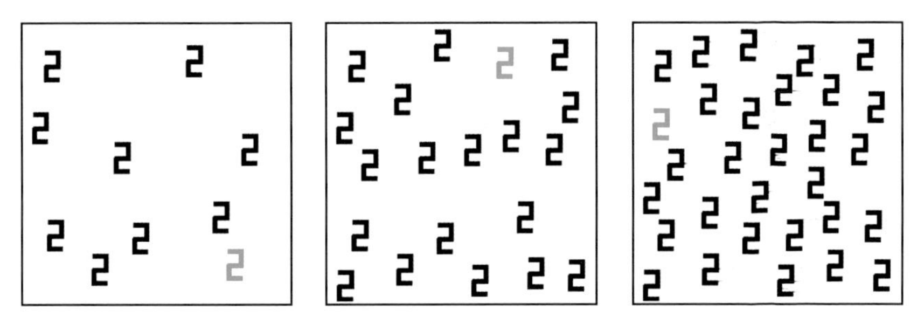

Fig. I: Color pop out. The green 2 is easily seen even when surrounded by many black 2s.

Fig. J: A store window display. This display makes it difficult to find brand or product information. © DONALD HOFFMAN

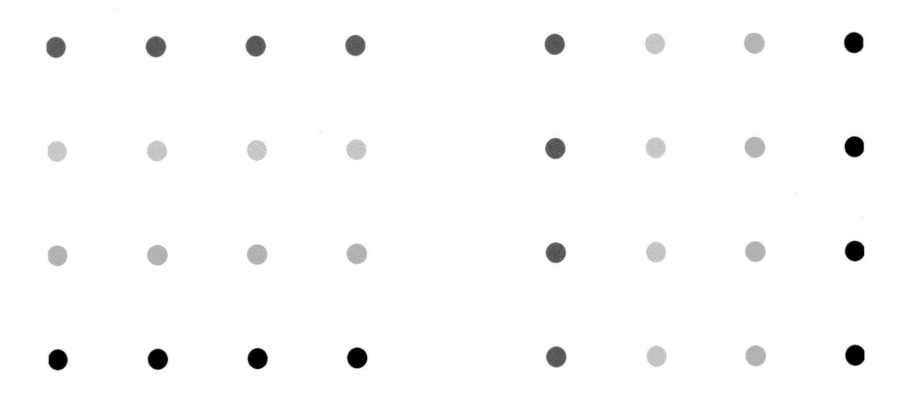

Fig. K: Grouping by color. We see horizontal groups on the left and vertical groups on the right.

ITP, which says that spacetime, as you perceive it, is like the desktop of an interface. If you look through a magnifying glass at the desktop of your computer, you'll see millions of pixels—the smallest patches of the desktop that are possible. Smaller than that, the desktop simply doesn't exist. Step back, and it looks like a continuous surface. If you play a video game on your computer, such as *Doom* or *Uncharted*, you see compelling 3D worlds with 3D objects. Yet the information is entirely 2D, limited by the number of pixels on the screen. The same is true when you look away from your computer to the world around you. It too has pixels, and all the information is 2D.

The physicists Leonard Susskind and Gerard 't Hooft helped to pioneer the holographic principle. Susskind says, "Here, then, is the conclusion that 't Hooft and I had reached: the three-dimensional world of ordinary experience—the universe filled with galaxies, stars, planets, houses, boulders, and people—is a hologram, an image of reality coded on a distant two-dimensional (2D) surface. This new law of physics, known as the holographic principle, asserts that everything inside a region of space can be described by bits of information restricted to the boundary."[34] This principle is now widely embraced in theoretical physics. Observers have no access to "objects" in "space." Observers *only* have access to information—bits—written on a boundary that *surrounds* space.

Black holes, which led to the holographic principle, have led another assault on our intuitions about spacetime. Hawking discovered that black holes radiate energy, now called Hawking radiation, whose temperature increases as the size of the black hole decreases. Hawking radiation takes energy out of a black hole, causing it to shrink and, eventually, evaporate altogether. Hawking claimed that, in this process, a black hole destroys all information about any objects that fall into it.[35] If a cat fell in, it would disappear into the black hole and all information about it would be forever annihilated.

That's bad for the cat, but also for quantum theory, which assumes that information is never eradicated. This is no minor assumption. If you take it away, quantum theory unravels into nonsense. Hawking's claim posed a serious threat.

Einstein's theory of general relativity says that a black hole sucks in and devours not just objects, but even space itself. As space gets sucked closer to the black hole, it flows faster, eventually reaching, and then exceeding, the speed of light. Nothing can travel through space faster than the speed of light. But that speed limit does not apply to space itself. Where space pours into the black hole at the speed of light, it is no longer possible for light, or information, to paddle upstream fast enough to escape. This is the event horizon of the black hole, the divide between the outside, where light can escape, and the inside, where escape is not possible.

According to Einstein, a cat falling through the event horizon would, if the black hole is big enough, experience nothing unusual. Eventually, as the cat plunged toward the center of the black hole, it would be "spaghettified," stretched beyond recognition by the rapidly changing force of gravity. But at the horizon it would just float on through, unaware that its fate was sealed.

According to Einstein, the cat and all its information coast across the event horizon never to be seen again. Then, when the black hole evaporates, so does all information about the cat.

Quantum theory says that information is never destroyed. General relativity says that it can cross an event horizon and be erased. This is a serious paradox.

It gets worse. Consider two cat lovers, Prudence and Folly. Prudence watches the cat at a safe distance from the black hole. She sees the cat approaching (but never passing) the event horizon, slowly stretching and deforming beyond recognition, and eventually getting barbecued by the Hawking radiation—a gruesome fate. Folly takes the plunge into the black hole with the cat. She sees something more pleasant—the cat passing safely through the horizon, with no contorting or torching. According to Prudence the cat and its information are mangled outside the horizon, but according to Folly the cat and its information are thriving inside the horizon.

But having the cat's information in two places—inside and outside the black hole—violates another rule of quantum theory: quantum information can't be copied. Not only is quantum information never destroyed, it can never be cloned. This is counterintuitive. I can copy information onto a hard drive. I can lose, or destroy, that drive. But my file consists of classical bits, which record classical information. Quantum information, however, is different from classical, and this raises the ante in the conflict between general relativity and quantum theory.[36]

Can we resolve this conflict without violating key principles of these pillars of science? The physicist Leonard Susskind found a way, using a concept from quantum theory: complementarity.[37] In classical physics you can specify an object's position and momentum at the same time. You can say that the instant after a soccer player kicks a ball its position on the field is this and its momentum toward the goal is that. But not in quantum physics. If you fire an electron out of an electron gun, you can precisely measure its position or its momentum, but not both at the same time. According to Heisenberg's uncertainty principle, the more you know about position the less you can know about momentum, and vice versa. The Kochen-Specker (KS) Theorem tells us, as we discussed earlier, that the position and momentum of the electron in fact have no real values independent of the kind of measurement—position or momentum—that you perform.

Susskind took complementarity to a new level, which he called "Black Hole Complementarity."[38] For the case of the cat, it says that the description of the cat inside the black hole is complementary to the description outside the black hole. You can observe a cat outside the horizon of the black hole being incinerated, *or* you can observe a flame-free cat inside the horizon. Both are legitimate, but complementary, descriptions. And here is the key point: no observer can see both descriptions of the cat, just as no observer can see the position and the momentum of an electron.

Susskind's idea is now called "horizon complementarity" because it applies not just to the horizon of a black hole, but to any event horizon, including the horizon that bounds the visible universe.

Horizon complementarity seems radical, but it works. It allows quantum theory and general relativity to coexist without contradiction. But we must let go of thinking we can describe spacetime and objects outside the horizon and, at the same time, inside the horizon. The assumption that we can see both, the assumption of a god's-eye view, which no observer can in fact take, is the problem. If we relinquish the divine view from nowhere, then quantum theory and general relativity can peacefully coexist. But the implications are stunning. One may dismiss the complementarity of an electron's position and momentum as an odd feature of tiny things. But this dismissal won't work for horizons of black holes. They can be millions of miles across. The vast spacetime inside a huge horizon is complementary to the vast spacetime outside. If we insist on a single objective spacetime that includes the inside and outside of a black hole—an idea embraced by Einstein and by common sense—then we put quantum theory and general relativity in conflict. If we let go of objective spacetime, then they enjoy a rapprochement.

Horizon complementarity challenges the idea that there is one objective spacetime that contains all observers. But physicists Joe Polchinski, Ahmed Almheiri, Donald Marolf, and James Sully (known by their last initials as AMPS) found another way to thrash this idea using quantum entanglement.[39] Again consider Folly and a black hole. But this time, let the black hole emit Hawking radiation until it shrinks to half its original size, at which point, quantum theory tells us, one can start to decode the information in the radiation.

According to quantum field theory, the vacuum is not just a big nothing. It is seething with pairs of virtual particles. Each member of an evanescent pair is entangled with its partner and has opposite properties. A pair appears and immediately their opposing properties annihilate each other, leaving a vacuum devoid of real particles. Now consider two such virtual particles, 1 and 2, that happen to appear right next to the horizon of the black hole and that, from Folly's view before she takes the plunge, don't obliterate each other. Instead, 2 falls into the black hole and 1 becomes, for her, a real particle of Hawking radiation.

Folly, before she jumps into the black hole, can measure that 1 is entangled with some particle, 3, in the Hawking radiation that emerged earlier from the black hole. She can then let herself slip into the black hole where she finds that 1 and 2 are entangled.

But this raises a problem: quantum theory requires entanglement to be monogamous. Particle 1 can be maximally correlated with particle 2 or particle 3, but not with both.

Horizon complementarity can't solve the AMPS problem because this problem is not about two observers separated by a horizon. It's about one observer, Folly, who sees 1 and 3 entangled and then sees 1 and 2 entangled. AMPS tried to solve the problem by proposing that there's a firewall at the horizon that incinerates poor Folly as she passes through, so that she never sees 1 and 2 entangled. This firewall rescues quantum theory, but it violates general relativity, which predicts that nothing unusual should happen at the horizon—Folly should coast through with no problem, and certainly shouldn't see a wall of fire suddenly appear from nothing.

The AMPS "firewall paradox" is causing consternation, and many efforts to resolve the paradox. Daniel Harlow and Patrick Hayden, for instance, discovered that it's not easy to decipher the Hawking radiation.[40] Using the best quantum computing possible, it would take Folly too much time to figure out that 1 and 3 were entangled. The black hole would already fizzle to nothing, so that Folly could not also observe that 1 and 2 were entangled. No observer can measure both entanglements.

Some physicists counsel avoidance of a god's-eye view by restricting physics to the "causal diamond" of an observer—the portion of spacetime that may interact with the observer.

For instance, the physicist Raphael Bousso proposes the principle of *observer complementarity*: "Each observer's experiments admit a consistent description, but a simultaneous description of both observers is inconsistent. This implies a fascinating conclusion which I will call observer complementarity. . . . Observer complementarity is the statement that a fundamental description of Nature need only describe experiments that are consistent with causality. . . . Observer complementarity implies that there must be a theory for every causal diamond, but not necessarily for spacetime regions that are contained in no causal diamond."[41]

The physicist Tom Banks, in an interview with science writer Amanda Gefter, makes a similar claim. "Relativity tells us that no observers are special. There has to be a gauge equivalence between causal diamonds, so everything outside my horizon is a gauge copy of the physics I can observe right here. So if you think of every possible causal diamond, you have an infinitely redundant description of the same quantum system seen by different observers . . . and spacetime emerges when you put all these descriptions together."[42]

This aligns with the claim of Fuchs, Mermin, and Schack, discussed earlier, "that reality differs from one agent to another. This is not as strange as it may sound. What is real for an agent rests entirely on what that agent experiences, and different agents have different experiences."[43] Quantum states vary from observer to observer. So does spacetime itself.

Which raises a perplexing question: What about the big bang? Didn't it happen 13 billion 799 million years ago, before any observers? Isn't it a fact of objective reality, not merely an interface description of an observer? If ITP says that spacetime is a feature of my desktop, not an insight into reality, then it says the same about the big bang. Surely no physicist would agree?

At least one physicist has argued that the universe has no history apart from observers, that "histories of the universe . . . depend on what is being observed, contrary to the usual idea that the universe has a unique, observer independent history."[44] That physicist was Stephen Hawking who, in collaboration with the physicist Thomas Hertog, favored a "top-down" cosmology that starts with the observer, rather than a "bottom-up" cosmology that assumes a god's-eye view.

They explain that "In our past there is an epoch of the early universe when quantum gravity was important. The remnants of this early phase are all around us. The central problem in cosmology is to understand why these remnants are what they are, and how the distinctive features of our universe emerged from the big bang."[45] Their point is that the colossal energy and density of a nascent universe demand a quantum mechanical description, with superpositions of states. The classical premise of a unique primeval state for the universe is inapt: "if one does adopt a bottom-up approach to cosmology, one is immediately led to an essentially classical framework, in which one loses all ability to explain cosmology's central question—why our universe is the way it is."[46]

So, although the move is radical, they abandon the bottom-up framework. "The framework we propose is thus more like a top-down approach to cosmology, where the histories of the universe depend on the precise question asked."[47] Measurements we make today—say, of the density of energy of the vacuum or of the rate of expansion of the universe—constrain the histories of the universe that we can entertain.

Hawking's cosmology concurs with Wheeler's experiment, discussed earlier, in which the billion-year history I ascribe to a photon from an ancient quasar depends on what I measure today. If I measure which path around a gravitational lens it took, then I am entitled to ascribe a billion-year history in which it went, say, through the top path. But I am not so entitled if, instead, I measure an interference pattern. Wheeler put it well. "Each elementary quantum phenomenon is an elementary act of 'fact creation.' That is incontestable. But is that the only mechanism needed to create all that is? Is what took place at the big bang the consequence of billions upon billions of these elementary processes, these elementary 'acts of observer-participancy,' these quantum phenomena? Have we had the mechanism of creation before our eyes all this time without recognizing the truth?"[48]

Hawking's approach coheres with the cosmology of QBism, in which quantum states are beliefs of observers, not scoops on reality. What I see now informs the states I assign to the past, including the big bang. As Fuchs says, "Noting how the Big Bang itself is a moment of creation with some resemblance to every individual quantum measurement, one starts to wonder whether even it 'might be on the inside.' Certainly QBism has creation going on all the time and everywhere; quantum measurement is just about an agent hitching a ride and partaking in that ubiquitous process."[49]

This chapter began with the prediction of ITP that spacetime and objects do not exist unperceived; they are not fundamental reality. I asked whether this prediction has been ruled out by physics in its quest for a theory of everything (TOE). We have a clear answer: it has not. Instead, it has remarkable support.

The brief tour of physics in this chapter is, to be sure, far from exhaustive. It omits interpretations of quantum theory—by Bohm, Everett, and others—that try to bestow reality on objects and spacetime.[50] My goal, however, was not a synopsis of physics, which would require a tome, but a brief on physics that shows ITP is not proscribed.

Remarkably, a key prediction of ITP—that spacetime must go before a TOE will come—is close to consensus among physicists. Nima Arkani-Hamed, for instance, in a 2014 lecture at the Perimeter Institute, mentions that "Almost all of us believe that spacetime doesn't exist, that spacetime is doomed, and has to be replaced by some more primitive building blocks."[51]

If spacetime is doomed, then so are its physical objects. They must be superceded by more primitive building blocks. But if spacetime is not the bedrock of reality, not the preexisting stage for the drama of life, then what is it? It is, I will suggest, a data-compressing and error-correcting code for fitness.

CHAPTER SEVEN

Virtuality

Inflating a Holoworld

"Many, many separate arguments, all very strong individually, suggest that the very notion of spacetime is not a fundamental one. Spacetime is doomed. There is no such thing as spacetime fundamentally in the actual underlying description of the laws of physics. That's very startling, because what physics is supposed to be about is describing things as they happen in space and time. So, if there's no spacetime, it's not clear what physics is about."

—NIMA ARKANI-HAMED, *CORNELL MESSENGER LECTURE 2016*

"There is no spoon."

—SPOON BOY, *THE MATRIX*

Science can demystify the exotic. This talent leads to new technology—from cell phones to satellites—which can seem, in the words of Arthur C. Clarke, "indistinguishable from magic."

Science can also mystify the mundane. It can plunge us without warning down a rabbit hole of the curious and curiouser. For instance, I see a spoon sitting *now* on the table over *there*. This is so comonplace that I'm not tempted to give it a moment's thought. But here, where I don't expect it, science injects a profound mystery: we still don't understand "now" and "there." That is, we don't understand time and space—length, width, and depth—which we take for granted, which are woven into the very fabric of our daily perceptions, and which we assume are a true and reliable guide to physical reality.

What we do understand, many physicists now tell us, is that spacetime is doomed. Space and time figure centrally in our daily perceptions. But even their sophisticated union into spacetime, forged by Einstein, cannot be part of a true description of the fundamental laws of nature. Spacetime, and all the objects it contains, will disappear in that true description. Nobel Laureate David Gross, for instance, observed, "Everyone in string theory is convinced . . . that spacetime is doomed. But we don't know what it's replaced by."[1] Fields medalist Edward Witten has also suggested that spacetime may be "doomed."[2] Nathan Seiberg of the Institute for Advanced Study at Princeton said, "I am almost certain that space and time are illusions. These are primitive notions that will be replaced by something more sophisticated."[3]

This is deeply unsettling. As Nima Arkani-Hamed explained, in the chapter's opening quote, "What physics is supposed to be about is describing things as they happen in space and time. So, if there's no spacetime, it's not clear what physics is about." For physicists this is wonderful news. To recognize a failure of a theory, no matter how dear that theory may be, is progress. Replacing the theory of spacetime with something more fundamental is an exciting challenge for creative theorists, and has the potential to transform our vision of the world—perhaps telling us, for the first time, what physics is really about.

My goal in this chapter is a tad less ambitious. The news that spacetime is doomed—and objects with it—does not yet inform current theories of visual perception. Instead, these theories typically assume that objects in space and time are fundamental in physical reality, and that visual perception normally recovers true properties of these preexisting objects. Current theories of perception often disagree about which true properties are reported, and about how the reports are generated, but they all assume to be true what physicists have discovered to be false—that objects in spacetime are fundamental.

I will briefly discuss the standard theories of perception, and then propose a new slant on our perception of spacetime and objects. The new perspective is motivated by ITP and the holographic principle—the momentous discovery, discussed in chapter six, that the amount of data you can store in a region of space depends on the *area* surrounding that region, not on its volume. This new outlook on spacetime and objects flows from the idea that our perceptions have evolved to encode fitness payoffs, and to guide adaptive behavior.[4] Somehow, spacetime and objects do just that. But how? I propose that they do it, in part, by data compression and error correction of fitness information.

First, let's look at data compression. A fitness-payoff function can be complex, and many fitness-payoff functions are typically relevant to my survival, so the amount of information about fitness that's pertinent to me could be enormous—overwhelming if I had to see it all. I need it compressed to a size I can manage.

Suppose you want to email a vacation photo to a friend, but the image is too large for your server. You compress the image and check that it still looks good. If it doesn't, if you can't see that it's your family posing by the Grand Canyon, then you compress it less. You look for a happy tradeoff—compressed enough to send, but not so compressed that it's not worth sending.

Spacetime and objects are, for human vision, that happy tradeoff. Fitness-payoff functions can vary in hundreds of dimensions. Human vision, shaped by eons of natural selection, compresses them into three dimensions of space and one dimension of time, and into objects with shapes and colors. I can't handle hundreds of dimensions, but I can handle a few. This compression no doubt omits some information about fitness. I don't, for instance, see the millions of muons that streak through my body each day, damaging it with ionizing radiation. But I do see enough information about fitness to survive and raise offspring.

We see objects in three dimensions not because we reconstruct objective reality, but because this is the format of a compression algorithm that evolution happened to build into us. Other species may have other data formats for representing fitness. We live and move and have our being not in an objective reality of spacetime and objects, but in a data structure with a format of spacetime and objects, which happened to evolve in *Homo sapiens* to represent fitness payoffs in a manner that is frugal and useful. Our perceptions are encoded in this data structure, but we mistakenly believe that its spacetime format is the objective reality in which we live. This mistake is understandable and even excusable: our data format constrains not just how we see, but how we think. It's not easy to step outside its confines, or even to recognize that this may be possible. Waking up to this possibility has a long pedigree in intellectual and religious culture.

There is much to explore about spacetime and objects as compressed encodings of fitness payoffs. For instance, what aspect of fitness is captured by space, and what by objects? How do shapes, colors, textures, and motions arise in the compression of fitness? Why does the compression of fitness lead us to have perceptions that are formatted in different modalities—vision, hearing, taste, smell, and touch? Perhaps distances in space encode costs of acquiring resources: an apple that costs few calories to acquire may appear just a meter away, while an apple that costs far more calories may appear much further away. A predator may appear more distant the more calories it must expend to get me. Recent experiments support this idea. For instance, Dennis Proffitt and his collaborators found that people given a drink containing glucose make shorter estimates of distance than those given a drink containing no carbohydrates (and, instead, an artificial sweetener); people who are more aerobically fit make shorter estimates of distance than those who are less fit. This suggests that our perception of a distance depends not just on the energy cost, but rather on the ratio of the energy cost to our available energy.[5]

Let's turn to error correction for a moment. When we bank or buy online, valuable data shoots across the internet. To prevent its theft by hackers, we encrypt it. But another problem is just as important: noise. Suppose you spend sixty dollars to buy flowers online for mom. Later you learn that noise on the net slipped two decimals, and you in fact spent six thousand dollars—an expensive mistake. If such mistakes were common, commerce online would halt. To prevent them, data are formatted in an error-correcting code before being sent.

A key to detecting and correcting errors is redundancy.[6] A simple example is repetition. Suppose that you want to send four bits of data, such as the bit string 1101. You could send it three times in succession: 1101 1101 1101. The receiver checks that all three transmissions agree. If so, then she concludes that there is no error. But if one transmission differs from the others, then she detects an error. She can ask for another transmission, or assume that the two strings that agree are correct.

There are many clever ways to add redundancy, such as embedding messages into higher-dimensional spaces. But the key point is that our senses convey messages about fitness payoffs, and getting the right message is critical to survival. Slip a decimal about fitness and you may slip from life to death. We should expect that natural selection has built redundancy into our perceptual interface, that it has shaped our desktop of spacetime and our icons of physical objects to be redundant codes for fitness payoffs that permit detection and correction of errors.

This is exactly what Bekenstein and Hawking discovered about spacetime. It is redundant. Two dimensions contain *all* the information in any 3D space. This is the well-established holographic principle of Susskind and 't Hooft that we discussed in the last chapter. It is counterintuitive, and belies our assumption that 3D space is an objective reality that our senses reconstruct. But it makes sense if you assume that our senses report fitness and need redundancy—such as an extra dimension of space—to ensure that their reports aren't crippled by noise.

Physicists have confirmed the prediction of natural selection that space is redundant. But have they also confirmed that in fact this redundancy of space underwrites an error-correcting code? That effort is under way and looks promising. The physicists Ahmed Almheiri, Xi Dong, and Daniel Harlow find that the redundancy of space revealed by the holographic principle reflects properties of an error-correcting code that protects against erasure of data by noise.[7] As they put it, "The holographic principle also naturally arises in the guise of the general statement that there is an upper bound on how much quantum information a given code can protect from erasures."[8] The physicists John Preskill, Daniel Harlow, Fernando Pastawski, and others have discovered specific ways that the geometry of spacetime can be interpreted as a quantum error-correcting code.[9]

The picture that emerges is that spacetime and objects are a code used by our senses to report fitness. Like any decent code, it uses redundancy to counter noise. This picture is precisely ITP, with the extra insight that the interface compresses data and resists noise.

This picture is not endorsed by most vision scientists. Instead, they assume that vision is veridical, that it reconstructs real objects in spacetime. This assumption is spelled out in the *Encyclopaedia Britannica* entry on "space perception" by Louis Jolyon West, former psychiatrist in chief at the UCLA Hospital and Clinics. West tells us in his entry that veridical perception is "the direct perception of stimuli as they exist. Without some degree of veridicality concerning physical space, one cannot seek food, flee from enemies, or even socialize. Veridical perception also causes a person to experience changing stimuli as if they were stable: even though the sensory image of an approaching tiger grows larger, for example, one tends to perceive that the animal's size remains unchanged."

Vision scientists don't claim, of course, that perception is always veridical. They admit that it can distort reality by using heuristics. But they assume that veridicality is the goal, and is normally attained.

They argue, for instance, that symmetries in our perceptions of objects reveal symmetries in objective reality. The vision scientist Zygmunt Pizlo spells this out. "Consider the shapes of animal bodies. Most, if not all of them, are mirror-symmetrical. How do we know that they are mirror-symmetrical? Because we see them as such. Seeing a mirror-symmetrical object as mirror-symmetrical is not possible unless the two symmetrical halves are perceived as having identical shapes. Now, note that this is remarkable because: (1) we only see the front, visible surfaces of each of the two halves, and (2) we see the two halves from viewing directions that are 180° apart. Unless shape constancy is a real phenomenon and unless it is close to perfect, we would not even know that symmetrical shapes actually exist."[10]

We can recast this as a precise claim: any symmetry in our perceptions entails a corresponding symmetry in objective reality.

Is this claim true? Here we don't need hunches, we need a theorem. And we have one. The "Invention of Symmetry Theorem," which I conjectured and Chetan Prakash proved, reveals that the claim is false.[11] This theorem states that symmetries in our perceptions entail nothing about the structure of objective reality. The proof is constructive. It shows precisely how perceptions and actions can enjoy a symmetry—such as translation, rotation, mirror, and Lorentz—in a world that lacks any symmetry.

This raises an obvious question. We see many objects with symmetries. Why? If symmetries in perception don't reveal symmetries of reality, then why should we see symmetry at all?

The answer, once again, is data compression and error correction—their algorithms and data structures often involve symmetries.[12] A surfeit of fitness information can be compressed to a feasible level using symmetries. To get a feel for this, consider looking at an apple. How will it look if you move a little to the left? You can answer this using symmetry—a simple rotation and translation. Rather than store millions of numbers per view, you need just five—three for translation and two for rotation. Symmetries are simple programs that we use to compress data and correct errors. The symmetries in our perceptions reveal how we compress and encode information, not the nature of objective reality.

"But," you might object, "we can build computer vision systems that drive cars and see the same shapes and symmetries that we do. Doesn't this demonstrate that we, and the computers, are seeing reality as it is?"

Not at all. The Invention of Symmetry Theorem applies to any perceptual system, whether biological or machine. The symmetries a computer sees entail nothing about the structure of objective reality. We can build a robot that sees the symmetries we see. But this grants us no insight into the structure of the world.

Pizlo offers an evolutionary rationale for veridical perceptions of objects and space. "It is not possible to conceive the successful evolution of animals, and the success of their natural selection without providing for planning and purposive behavior."[13] He argues that our success in hunting, planting, and harvesting depends on planning and coordination, which require veridical perception of objective reality.

Planning and coordination are critical to our success. But do they require a veridical representation of objective reality? No, according to the FBT Theorem. Indeed, online games such as *Grand Theft Auto* let players collaborate toward ignoble goals, such as robbing stores or stealing cars. Their plans are informed not by veridical perceptions of transistors and network protocols, but by a fake world of fast cars and tempting targets.

The arguments for veridical perception fail. But it is still the standard theory in vision science. According to this theory, there really are 3D objects in spacetime with objective properties—such as shape —that exist even when no one looks. When you look at an apple, light bouncing off its surface gets focused by the optics of your eye onto your 2D retina. This optical projection of the apple onto your 2D retina loses information about the apple's 3D shape and depth. So your visual system analyzes its 2D information and figures out the apple's true 3D shape. It recovers, or reconstructs, the information lost by the optical projection. Sometimes this recovery process is called "inverse optics," and sometimes "Bayesian estimation."[14]

Proponents of "embodied cognition," building on the ideas of the psychologist James Gibson, push back on this story.[15] They say that we are physical beings with real bodies that interact with the real physical world, and that our perceptions are intimately linked with our actions. Perception and bodily action must be understood together. When I see a red apple, I am not simply solving an abstract problem of inverse optics or Bayesian estimation, I see a 3D shape that is tightly coupled to my actions—how I move toward it, grasp it, and eat it. Most vision scientists who subscribe to inverse optics or Bayesian estimation agree that action and perception are intimately linked.

Proponents of "radical embodied cognition" claim not just that perception and action are linked, but also that perception requires no processing of information.[16] The interplay of perception and action can be understood, they claim, without invoking computations and representations. This radical view has few devotees and is at odds with the claim of quantum physicists that all physical processes are information processes, and that no information is ever destroyed. It is also at odds with the truism that any system that undergoes a sequence of state transitions can be interpreted as a computer (perhaps a dumb one, but a computer nonetheless).

ITP disagrees with the claim of standard and embodied theories that perception is veridical, but it agrees that perception and action are closely linked. Our perceptions evolved to guide adaptive exploration and action: my icon of an apple guides my choice of whether to eat, as well as the grasping and biting actions by which I eat; my icon of poison ivy guides my choice not to eat, as well as the steps I take to avoid any contact.

ITP makes a counterintuitive claim about causality: the appearance of causal interactions between physical objects in spacetime is a fiction—a useful fiction, but a fiction nonetheless. I see a cue ball hit an eight ball into a corner pocket. I assume, naturally, that the cue ball caused the eight ball to careen to the corner. But strictly speaking, I'm wrong. Spacetime is simply a species-specific desktop, and physical objects are icons on the desktop; or, as we have just been discussing, spacetime is a communications channel and physical objects are messages about fitness. If I drag an icon to the trashcan and its file gets deleted, it's often helpful, though mistaken, for me to think that the movement of the icon to the trashcan literally caused the file to be deleted. Indeed, the ability to predict the consequences of one's actions through this kind of pseudo, cause-effect reasoning is a sign of a well-designed interface.

This prediction of ITP—that the appearance of causal interactions between physical objects in spacetime is a fiction—has interesting support from quantum computations that lack causal order.[17] Normally we compute one step at a time, in a specific causal order. I might, for instance, start with the number ten, divide it by two, and then add two, to get the result seven. If I reverse the order, if I add two and then divide by two, I get the result six. The order of operations matters. But computers have now been built in which there is no definite causal order of operations. Instead the computer uses a superposition of causal orders, resulting in more efficient computation.[18]

The interface theory predicts that physical causality is a fiction. This is not contradicted by physics. If, as physicists now say, spacetime is doomed, then so also are its physical objects and their apparent causality. So are current theories of consciousness, such as the integrated information theory (IIT) of Giulio Tononi or the biological naturalism of John Searle, that identify consciousness with certain causal properties of physical systems in spacetime.[19] If physical objects such as neurons have no causal powers, then IIT identifies consciousness with a fiction—not a promising move. Moreover, causal computations are less powerful than computations that abandon causality.[20] When IIT identifies consciousness with causal computations, it identifies consciousness with inferior computations. Why should consciousness be inferior? What principled insight about consciousness dictates this dubious claim?

The fictive nature of physical causality makes it tricky to construct the elusive "theory of everything." We must first postulate a theory of our interface, and of its various levels of data compression and error correction. Then we can use this theory to ask what, if anything, we can infer about objective reality from the structures we see in the interface. If we can't infer anything, then we must postulate a theory of objective reality and predict how it appears in our interface. This is the normal scientific process of using our theories to make empirical predictions that can be tested by careful experiments. I suspect that, if we succeed in this enterprise, we will find that the distinction we make between the living and nonliving is an artifact of limitations of our spacetime interface, not an insight into the nature of reality. We will find a unified description for reality—animate and inanimate—once we take into account the limits of our interface. We will also find that networks of neurons are among our symbols for error-correcting coders.

In ITP we can visualize the link between perception and action in a simple diagram, shown in Figure 10, in which an agent interacts with the world. The rounded box at the top of the diagram represents the world outside the agent. I won't claim, for now, to know anything about this world. In particular, I won't assume that it has space, time, or objects. I'll simply say that this mysterious world has many states—whatever they may be—that can change. The agent, for its part, has a repertoire of experiences and actions, shown in rounded boxes. Based on its current experience, the agent decides whether, and how, to change its current choice of action. This decision is depicted by the arrow labeled "decide." The agent then acts on the world, as depicted by the arrow labeled "act." The action of the agent changes the state of the world. The world, in response, changes the experience of the agent, as depicted by the arrow labeled "perceive." Perception and action are thus linked in a "perceive-decide-act" (PDA) loop (which is described mathematically in the appendix).

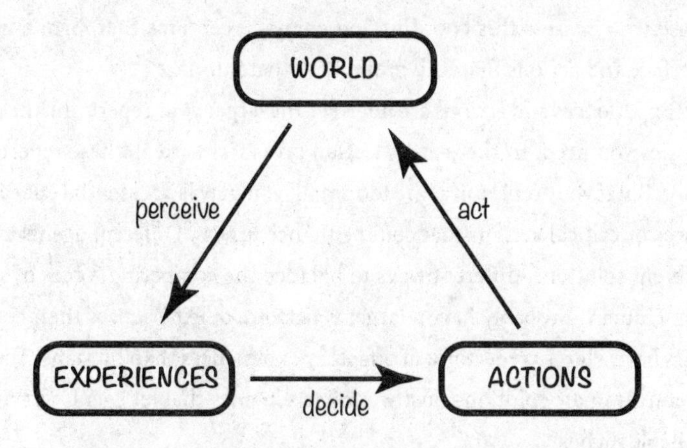

Fig. 10: The "perceive-decide-act" (PDA) loop. Natural selection shapes this loop so that experiences guide actions that enhance fitness. © DONALD HOFFMAN

The PDA loop is shaped by an essential feature of evolution—the fitness-payoff functions. The fitness of an action depends on the state of the world, but also on the organism (the agent) and its state. Each time an agent acts on the world, it changes the state of the world, and reaps a fitness reward (or punishment). Only an agent that acts in ways that reap enough fitness rewards will survive and reproduce. Natural selection favors agents with PDA loops properly tuned to fitness. For such an agent, its "perceive" arrow sends it messages about fitness, and its experiences represent these messages about fitness. The messages and experiences are all about fitness, not about the state of the world. The experiences of the agent become an interface—not perfect, but good enough. It guides actions that glean enough fitness points to survive long enough to rear offspring.

Each agent has been molded, through generations of ruthless selection, to decide on actions that lead to desirable payoffs in fitness. The reproductive imperative, that one must act in ways that collect enough fitness points to raise offspring, coerces the coordination of perception, decision, and action. Those who lack this coordination suffer a pathetic proclivity to die young. Those who possess this coordination enjoy perceptions that form a useful interface and actions that link properly to that interface.

Experiences and actions are not free. The larger your repertoire, the more calories you need, so there are selection pressures to keep these repertoires small. But if your repertoires are too small, you may lack essential data about fitness or critical actions that could enhance fitness. Different agents evolve different solutions, different ways to balance the competing forces of selection. Humans probably have a larger repertoire of experiences than beetles; bears have a larger repertoire of olfactory experiences than humans. There is no consummate solution—just workable schemes that let agents survive in available niches.

But in all solutions, the repertoire of experiences and actions is small compared to the complexity of the relevant fitness payoffs. All messages about fitness that an agent perceives must compress information about fitness into a manageable size and useful format, without losing critical information. And messages should allow an agent to find and correct errors.

For instance, you're strolling along a sidewalk at dusk, and suddenly jump in fear. You peer around to find a culprit, and relax when you discern a garden hose in the grass. Your jump was triggered by a fitness message with inadequate error correction—it incorrectly said "snake." Because this message didn't waste time on error correction, it arrived quickly and you acted promptly to avoid a fitness-reducing bite. After your initial startle, an error-corrected message arrived saying, "No worries, just a hose." Your needless jump wasted calories and triggered stress-inducing cortisol, so it slightly pared your fitness. But in the long run, such quick and fallible messages stoke your fitness by slashing the risk of a mortal bite. If you trafficked only in plodding but reliable messages, then you would hasten the day that you correctly learn, "You've just been bitten." Correct, but less helpful.

This illustrates that there are multiple solutions to the problem of compressing and correcting fitness messages. We can expect that natural selection has shaped a variety of solutions tailored to the vagaries of fitness, and that a single organism may embody multiple solutions for its different fitness needs. But we can also expect to find similar solutions across species because evolution, in the process of speciation, will often repurpose rather than redesign. We see repurposing in the unintelligent design of our eyes: light that passes through the lens of the eye must negotiate a gauntlet of blood vessels and interneurons before it chances on a photoreceptor at the back of the retina. All vertebrates suffer this kludge, suggesting that it cropped up early in vertebrate evolution and was never corrected. The kludge isn't necessary. Cephalopods, such as the octopus and squid, get things right—their photoreceptors sit in front of the interneurons and blood vessels.

We can see error correction in real time in the visual example shown in Figure 11. On the left are two black disks with white cutouts. On the right these disks are rotated so that their cutouts align. Suddenly you see more than disks with cutouts. You see a glowing line that floats in front of the disks. You can check that you create the glow between the disks: cover the disks with your thumbs, and the glow disappears.

You can think of the glowing line as your correction of an erasure. It's as though your visual system decides that the actual message that was sent was a straight line, but that part of the line got erased in transmission. It corrects the error by filling in the gap with a glowing line. This is similar to error correction in a simple "Hamming" code that can send only two messages: 000 or 111.[21] If the receiver gets, say, 101, then it knows that there was an error, that

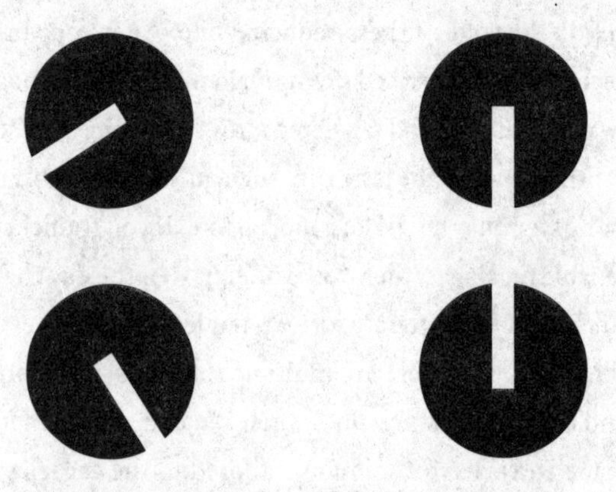

Fig. 11: Correcting an erased line. The visual system creates a line between the two disks on the right to correct an erasure error. © DONALD HOFFMAN

the middle 1 got erased, so it fixes the erasure and arrives at the message 111. This Hamming code uses three bits to send just one bit of information, so it allows the receiver to detect and correct one erasure error.

By correcting the erasure in the image of black disks you recover a message: "line in front of disks." You can also recover a second message: "line behind disks." To see this message, think of the disks as holes in a sheet of white paper. You're looking through the holes, and behind the paper you see a line. Notice that when you see this line, the segment of the line between the disks no longer glows, but you still sense that it's there.

Which line is there—glowing, or not glowing—when you don't look? The question is of course silly. There is no line when you don't look. Instead, the line you see is the message you recover when you correct an erasure.

Let's ask a different question: Which line *will* you see—glowing, or not—when you look? You can't be certain. Sometimes you'll see a line that glows, sometimes a line that does not. But you can guess probabilities. I see the glowing line more often. I would say that the probability is about three-quarters that I will see it glowing and one-quarter that I will see it not glowing. If someone demanded that I write down my probabilities in terms of the "states" of the line—glowing, or not glowing—then I would write down a "superposition" state for the line, in which the glowing state has a three-quarters probability and the not-glowing state has a one-quarter probability. This is analogous to the superposition of states that we encountered earlier in quantum theory. Recall that, according to QBism, a quantum state does not describe the objective state of a world that exists even if no one looks, but rather it describes the beliefs of an agent about what she will see if she acts, or, to put it more technically, what outcome she will obtain if she makes a measurement.[22]

Let's take this example a step further. In Figure 12 there are, on the left side, four black disks with white cutouts. On the right these same disks are rotated so that their cutouts align. Suddenly you see more than disks with cutouts. You see four glowing lines that float in front of the disks. Each glowing line seems to continue through the blank space between disks. You can again check that you are creating the glow between disks by covering up two disks with your thumbs; the glow disappears.

Fig. 12: Correcting an erased square. The visual system creates a square over the four disks on the right to correct an erasure error. © DONALD HOFFMAN

Your visual system has corrected four erasure errors and created four glowing lines. But it also detects another coded message, at yet a higher level: it detects a square. It receives messages at different levels of abstraction—one-dimensional lines and a 2D square. Your correction of errors probably involves both levels at once; the evidence that the message is a square increases the confidence of your visual system in the evidence that lines were erased and should be restored.

Your visual system can detect a second message about a square. Again, think of the four black disks as holes in a white sheet of paper, and imagine that you're looking through these holes. Then behind the paper you'll see a square. When you do, notice that its lines don't glow. You're confident that the lines are there, but they're hidden by the white paper.

So you can get two different messages about a square from this figure. One message has the square in front, with glowing lines; the second message has the square in back, with lines that don't glow. Notice that all four lines glow, or else all four lines do not glow. You never see, say, two lines glowing and two not glowing. Why? Because your visual system has united all four lines into a single unified message: a square. It has "entangled" the four lines into a single object so that what happens to one line must happen to all.

Now let's take our example one final step. In Figure 13 there are, on the left, eight black disks with white cutouts. On the right these same disks are rotated so that their cutouts align. Suddenly you see twelve glowing lines; you have corrected twelve erasures of lines.

But now you do something radical: you entangle these lines to form a single

Fig. 13: Correcting an erased cube. The visual system creates a cube over the eight disks on the right to correct an erasure error. ADAPTATION OF THE SUBJECTIVE NECKER CUBE. BRADLEY, D. R. AND PETRY, H.M. "ORGANIZATIONAL DETERMINANTS OF SUBJECTIVE CONTOUR. THE SUBJECTIVE NECKER CUBE." *AMERICAN JOURNAL OF PSYCHOLOGY*, 1977, 90, 253-62 (SEE NOTE 23).

object—a cube—and, in the process, you create a new dimension of depth.[23] You start with information in two dimensions and then inflate it, holographically, into three dimensions. Entanglement in this example is intimately linked with the creation of a conscious experience of three dimensions of space. Notice that sometimes you see a cube with corner A in front and other times you see one with corner B in front. When you flip from one cube to the other, you reverse the relationships of depth in three dimensions that you holographically construct—lines that were in front go to the back, and vice versa. That the lines are all entangled can again be verified by noting, for instance, that they all glow when the cube is seen in front of the disks and they all cease to glow when the cube is seen as behind the disks. This cube, known as the "subjective Necker cube," was originally described by Heywood Petry, and is widely recognized by vision scientists as an inspired illustration and demonstration of the human visual capacity to create subjective contours and objects in three dimensions.

In quantum theory, work by Mark van Raamsdonk, Brian Swingle, and others indicates that spacetime is woven together from threads of entanglement.[24] I suspect that there is more than mere analogy here. I suspect that superposition, entanglement, and the holographic inflation of three dimensions seen in our visual example is precisely the same as studied in quantum theory. Spacetime is not an objective reality independent of any observer. It is an interface shaped by natural selection to convey messages about fitness. In the

Fig. 14: Shaded disks. The random shading of the left disk and the uniform shading of the middle disk makes them look flat. The shading of the right disk makes it look like a sphere. © DONALD HOFFMAN

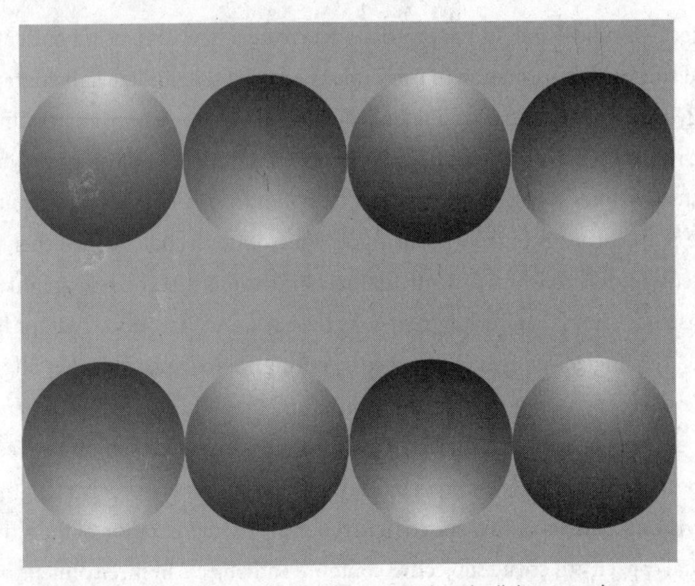

Fig. 15: Convex and concave disks. We assume that the light source is overhead. © DONALD HOFFMAN

visual example of the cube we see this spacetime interface in action, complete with error correction, superposition, entanglement, and holographic inflation.

Another way you inflate two dimensions into three is shown in Figure 14. On the left is a disk in which the brightness of each point is chosen at random. You just see noise. In the middle is a disk of constant brightness, which looks flat. But on the right is a disk in which brightness varies gradually and systematically. Now the magic happens—you inflate the disk into a sphere. Even though the information is 2D, you holographically inflate it into a 3D object.

Sometimes, as shown in Figure 15, you inflate a shape that is convex, and other times you inflate one that is concave: your visual system prefers to inflate a shape in such a way that it appears to be lit from overhead.[25]

In addition to inflating gradients of brightness, you also inflate curves, as shown in Figure 16. On the left is a disk with a grid of straight lines, which looks flat. In the middle, the lines are curved slightly, and you inflate a sphere. On the right, curved lines and gradients of brightness are combined, and you inflate a compelling sphere.

Fig. 16: Inflating the third dimension. We sometimes interpret curving contours as a shape with depth in three dimensions. © DONALD HOFFMAN

What do we learn from these examples of lines, squares, cubes, and spheres? According to standard vision science, they show us how the visual system reconstructs the true shapes of real objects in an objective spacetime.

According to ITP, they show us something entirely different—how the visual system decodes messages about fitness. There is no objective spacetime and no preexisting objects in spacetime whose true properties we try to recover. Instead, spacetime and objects are simply a coding system for messages about fitness. The visual examples we have just seen, in which we catch ourselves inflating information from two dimensions into three, don't show that objective reality has two dimensions rather than three. Instead, they are intended to weaken our conviction that spacetime itself is an aspect of objective reality. The examples have two dimensions simply to fit on the page.

If a fitness message is corrupted by a little noise, then the system can sometimes correct the error, as we saw with the glowing lines. If the noise is too great, as in the disk whose pixels have random brightnesses, then we cannot correct the error; we see noise with no clear fitness message.

But if brightness and contours convey a consistent message, then we often decode that message into a language of 3D shapes that is tailored to guide adaptive action. We see, for instance, a sphere and thereby know how to grasp it or avoid it. We see an apple and know that grasping and eating it can enhance our fitness; we see a leopard and know that the same actions are unwise.

In short, we do not recover the true shape in three dimensions of a preexisting object—there are no such objects. Instead, we recover a message about fitness that happens to use shapes in three dimensions as a coding language.

Once we know the rules that human vision uses to decode messages about fitness, we can use those rules to send the messages we want. Consider jeans. They often have finishes, sanded by hand or etched by a laser, that are intended to mimic wear and tear. These finishes have brightness gradients, like the brightness gradient of the sphere in Figure 16, that convey a message about a shape in three dimensions. Jeans also have curved contours—pockets, seams, and yokes. Like the curves of the sphere in Figure 16, these

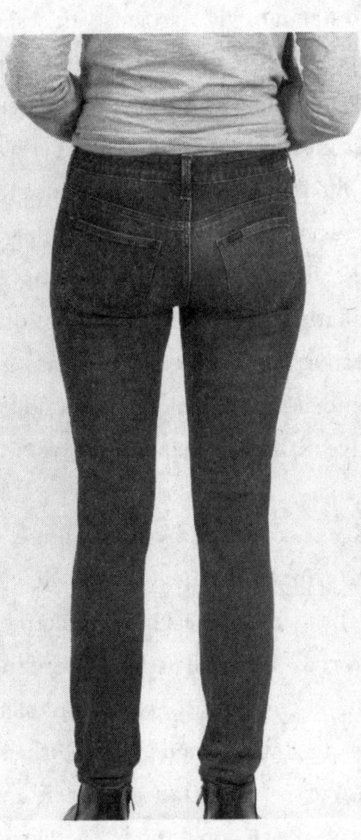

Fig. 17: Enhancing the body with jeans. The left side looks flat. The right side looks firm and toned. The difference is due to careful use of visual cues for depth.

convey a message about a shape in three dimensions. Darren Peshek and I found that by carefully arranging these curves and finishes, we could alter the perceived shape to convey another message about fitness—that the body wearing the jeans is attractive. This led to a new line of clothing known as Body Optix™.[26] Clothing, like makeup, can send carefully crafted messages—with a few white lies—about fitness.

This is illustrated by the pair of jeans in Figure 17 (this image can be viewed in full color in the Color Insert as Figure A). On its left side, the jean has a standard construction and finish. On its right side, it has a construction and finish carefully designed to convey the message of a well-toned and attractive body. The left side looks flat; the right side, shapely and toned. One person wears the jeans, but their two sides differ sharply in apparent shape and attractiveness.

In sum, spacetime is not an ancient theater erected long before any stirrings of life. It is a data structure that we create now to track and capture fitness payoffs. Physical objects such as pears and planets are not antique stage props in place long before consciousness took the stage. They too are data structures of our making. The shape of a pear is a code that describes fitness payoffs and suggests actions I might take to ingest them. Its distance codes my energy costs to reach it and snatch it.

We inflate spacetime and construct objects with carefully crafted shapes. But then we add a flourish. We paint these shapes with colors and textures. Why? Because colors and textures code critical data on fitness, as we will explore in the next chapter.

CHAPTER EIGHT

Polychromy

Mutations of an Interface

"Mere color, unspoiled by meaning, and unallied with definite form, can speak to the soul in a thousand different ways."

—OSCAR WILDE, *THE CRITIC AS ARTIST*

Color can speak volumes. It can direct a thousand different messages about fitness payoffs, and trigger, for each, an adaptive response. Color is a window on fitness—and also a jailhouse. Try to imagine a specific color that you've never seen. I've tried, and nothing happens. Surely there are colors that other people, or other animals, have seen that I have not, but I cannot concretely imagine even one of them, just as I cannot visualize a space having four dimensions. Color, like each of our perceptions, is both window and prison.

As a window on fitness, color is not flawless, just adequate to guide actions that keep us alive long enough to reproduce. Color, like each of our perceptions, compresses the complexities of fitness payoffs to bare essentials.

Every window has a bounding frame. The human eye only sees light with wavelengths between about four hundred and seven hundred nanometers—a minuscule fraction of the entire electromagnetic spectrum. This is not just data compression, it is data deletion. Outside our tiny window of color there are volumes of data about fitness, which we discard at our peril, including microwaves that can cook us, ultraviolet rays that can burn us, and X-rays that can give us cancer. What we don't see can, and sometimes does, kill us. But it usually does so only after we've raised offspring. So, to these perils that rarely impair our chance to reproduce, natural selection leaves us blind and vulnerable. Our perceptions tell us about fitness, but what they say is not veridical or unabridged. They tell us less than we may selfishly wish for—enough to have and raise children, but not enough to make us vibrant centenarians.

There is a wealth of information within the tiny window of wavelengths that we can see. Yet we compress it ruthlessly, down to just four numbers at each tiny region of the eye. We get three of the numbers from photoreceptors called cones, which come in three kinds—L, M, and S—and the last number from photoreceptors called rods.[1] The way they compress data is illustrated in Figure 18 (this image can be viewed in full color in the Color Insert as Figure B).

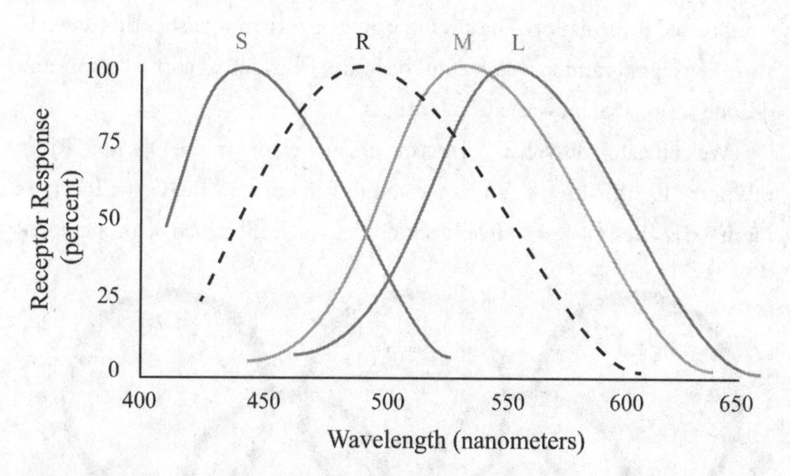

Fig. 18: Sensitivity curves for the three types of cones in the retina of the eye (L, M, and S). The sensitivity of rods, which mediate vision in low light, is given by the "R" curve.

Consider the red curve labeled "L." It shows the sensitivity of the L cone to various wavelengths of light. If a photon of light has a wavelength of about five hundred and sixty nanometers—near the top of the red curve—then the L cone has a much better chance of catching it and sending a signal than if a photon has a wavelength of 460 nanometers—near the bottom of the red curve.

Similarly, the M cone is most sensitive to light at about 530 nanometers, and the S cone is most sensitive at about 420 nanometers. These three cones—L, M, and S—are critical to our perception of color and are most useful in bright light. The remaining dashed curve, labeled R, shows the sensitivity of rods, which mediate our vision of shades of gray in dim light. The overall sensitivity of the rods is much higher than that of cones, allowing them to operate in dim light.

This is massive compression of data. We ignore all photons outside a minuscule window of wavelengths and squeeze the remaining sliver of photons through the four filters of Figure 18.

The human eye has 7 million cones and 120 million rods, each carrying compressed information. The circuitry of the eye then squashes this down to 1 million signals and forwards it to the brain, which must correct errors and decode actionable messages about fitness.

We can catch ourselves correcting erasure errors in the Olympic-Rings of Figure 19 (this image can be viewed in full color in the Color Insert as Figure C). The image has five black circles, each inscribed with a colored

Fig. 19: The Olympic rings illusion. The colors that fill each ring are illusory. The visual system creates them to correct an erasure error. © DONALD HOFFMAN

circle. The interior of this circle is white. Your visual system detects an error. It presumes that the inscribed color once filled the disk but got erased. It fixes the erasure by injecting color. You see faint disks of blue, orange, gray, green, and red. The effect is strongest if you look slightly to the side of the figure. This "watercolor illusion" was exploited in older maps of the world to paint countries with distinct colors.[2]

We can catch ourselves again in the act of correcting color errors in the neon-square illusion shown in Figure 20 (this image can be viewed in full color in the Color Insert as Figure D).[3] The image on the left consists of black circles with arcs painted blue. The space between circles is white. But your visual system presumes that a transparent blue square was erased, and it corrects the error by filling in a glowing blue square with sharp edges. You can check that the square is illusory by covering the circles; the blue glow disappears.

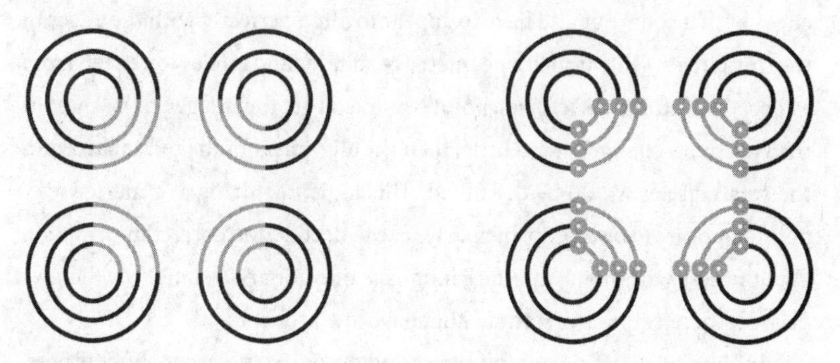

Fig. 20: The neon square illusion. The glowing blue square is illusory. The visual system creates it to correct an erasure error. © DONALD HOFFMAN

Your correction of errors and decoding of color follows a sophisticated logic that vision scientists are still working to understand. The right side of Figure 20 is just like the left side, except that little blue circles have been added. Although the image on the right has more blue contours than the image on the left, you no longer presume that a blue square was erased, and you no longer paint in a glowing square.

Your logic here appears to involve sophisticated reasoning about

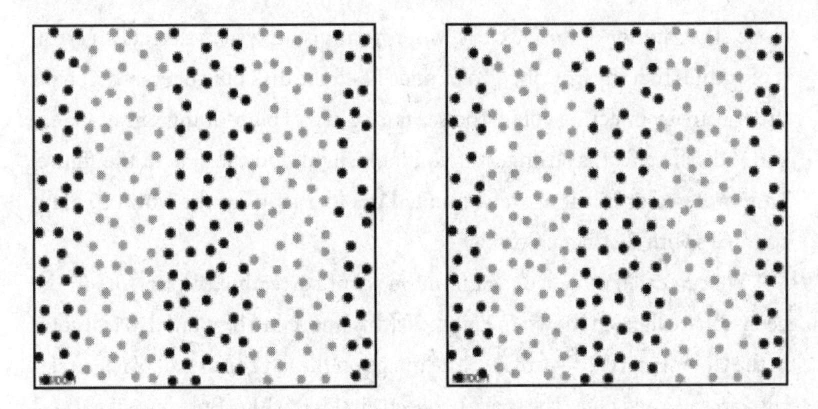

Fig. 21: Two frames of dots from a movie. When the frames are displayed as a movie, the visual system creates blue bars that move, glow, and have sharp edges. © DONALD HOFFMAN

geometry and probability. If a red transparent square were floating just a tad above a pattern of large and small circles in the image on the right, then the edges of that square would have to appear to align perfectly with the edges of the tiny circles. Only if such a geometry of squares and circles were seen from a special, or "nongeneric," viewpoint would you get the image on the right. If the viewpoint changed just a little, then the alignment of the red square with the small circles would be disrupted. This logic, requiring a "generic viewpoint," appears to be a key principle we use to decode and correct information about fitness within our interface language of color and geometry; when we decode, we reject interpretations that have low probability.[4]

In the process of correcting errors and decoding messages about fitness, we sometimes construct complex icons that integrate objects, colors, and motions. Figure 21, for instance, shows two frames from a movie available online (this image can be viewed in full color in the Color Insert as Figure E).[5] Each frame contains dozens of dots, each dot keeping its same position from frame to frame. From one frame to the next, some dots change color, either from black to blue or vice versa. But when you view the movie, you see blue bars with sharp edges scrolling to the left over a field of black dots.[6] You fill the white space between blue dots with a transparent blue surface, correcting an erasure. You delimit this blue surface with sharp edges, correcting another

Fig. 22: Joseph's hat illusion. The brown rectangle on the left side of the hat is printed in the same color ink as the yellow rectangle on the front of the hat. © DONALD HOFFMAN

erasure. You bind the edges and the blue surface to create a single object, a transparent bar, and then attribute a leftward motion to your creation. You have, by the end of this process, decoded a message about fitness into the language of your interface—the language of objects with shapes, positions, colors, and motions—a message that can now guide your next action.

Complex shapes guide complex actions. Consider Joseph's hat in Figure 22 (this image can be viewed in full color in the Color Insert as Figure F). You decode complex shapes for its brim and crown, which undulate in three dimensions. As a result, you know that to grasp it by the brim requires your hand to adopt certain grips and orientations, whereas to grasp it by the crown requires others. You know that your hand can grasp the brim more firmly than the crown without distorting its shape. The hat is an icon of your interface whose complex shape encodes information critical to adaptive action.

Your hand itself is an icon of your interface, not an objective reality. You must decode the shape of your hand, no less than that of the hat. We don't know what the objective world really is, and so we don't know exactly what we're really doing in that objective world when we grasp a hat. All we know is that, whatever we're really doing, our interface only lets us see a 3D hand grasping a 3D hat. Hat and hand, and grasping hat in hand, are messages about fitness that are compressed and coded in the error-correcting format that we perceive as 3D space. My very body is an icon, hiding a complex reality of which I'm ignorant. I don't know my real actions. I know only how the icon of my body appears to interact with other icons in my interface.

Joseph's hat sports many colors, which we decode as surfaces and lights. We interpret the brown rectangle on the left side of the hat as a brown surface in direct light, and the yellow rectangle on the front of the hat as a yellow surface in shadow. You can also see these two rectangles as the same color: if you cover all of the hat except these rectangles, then they look the same brown. (In fact, when creating this image, I used the dropper and paint-bucket tools of Photoshop to make the pixels in the two rectangles identical.) You can decode this image in two conflicting ways, one in which the rectangles are the same brown, and one in which they have different colors. Neither portrays objective reality. Both are simply messages about fitness. You decode disparate messages in different contexts.

The hat is an icon whose shapes and colors help you to secure fitness payoffs. Its description is not exhaustive, just what you need in the moment. Its shape informs you how to grasp it, and how to place it on your head to best protect you from the elements. It also has a category—hat—that offers useful tips about fitness: hats don't bite, they're inedible, they don't run, but they do protect from sun and cold. An icon of a different category—say, snake—offers different tips: it bites, it's edible, it doesn't run but does slither quickly, and it won't protect you from the weather. If you are forced to grasp it, its shape informs you to use a different grip than you would use with a hat.

As we have discussed, the idea that physical objects are just ephemeral data structures that describe fitness payoffs differs sharply from the idea—now standard in vision science—that physical objects are elements of objective reality, and that the goal of vision is to estimate their true shapes and other physical properties. It also differs from the claim that our interactions with physical objects give us direct, noninferential access to their real properties.

These differences are basic. The interface theory says that space and time are not fundamental aspects of objective reality, but simply a data format for messages about fitness, a format evolved to compress and correct such messages. Objects in spacetime are not aspects of objective reality, but simply messages about fitness coded in a format of icons that is specific to the needs of *Homo sapiens*. In particular, our bodies are not aspects of objective reality, and our actions don't give us direct access to preexisting objects in spacetime. Our bodies are messages about fitness that are coded as icons in a format specific to our species. When you perceive yourself sitting inside space and enduring through time, you're actually seeing yourself as an icon inside your own data structure.

Our senses evolved to encode fitness payoffs in a language of experiences. That language includes our experience of emotions. From anger, fear, distrust, and hate to love, joy, peace, and bliss, our emotions comprise a rich vocabulary. Specific emotions may be triggered by specific colors, a possibility now being studied by the science of color psychology.[7] Preliminary results suggest the following associations:

red lust, power, hunger, or excitement;
yellow jealousy or happiness;
orange comfort, warmth, or fun;
green envy, harmony, or good taste;
blue competence, quality, or masculinity;
pink sincerity, sophistication, or femininity;
purple power or authority;
brown ruggedness;
black grief, fear, sophistication, or expensiveness;
white purity, sincerity, or happiness.

This list paints with a wide brush. There are, for instance, many shades of red, each with its unique hue, saturation, and brightness. Fire-engine red feels nothing like a burgundy; the emotion evoked by a color surely depends on its specific shade.

The evoked emotion also depends on visual context. The patch of brown on the left side of Joseph's hat in Figure 22 (Color Insert F) has the hue and saturation of "opaque couché"—a greenish brown voted by thousands of Australians to be the ugliest color in the world. The same patch on the front of the hat looks yellow, which is not the ugliest color in the world. The pixels in both patches have the same color coordinates. But these color coordinates evoke different emotional responses in the two different visual contexts.

The evoked emotion may depend on culture: the shade of red ubiquitous in the regalia of Spanish bullfights may signify emotions to Spaniards, such as exciting danger or national pride, which would be lost on most Americans. The emotion may depend on particularities of personal experience: the shade of yellow flourished by banana spiders may evoke idiosyncratic fears in some arachnophobes.

Nuances of color can trigger nuances of emotion that inform our actions in pursuit of fitness. Even plants, which may have no emotions, use nuances of color to guide a variety of adaptive actions. The growth tips of some plants have photoreceptors that detect blue light and guide growth toward open sky.[8] They hunt light much as we hunt game, tracking blue photons to wrangle light.

The leaves of some plants have photoreceptors that are sensitive to red light. When they catch red light, the plant "knows" that it's morning, and when they subsequently catch a deeper red light, the plant knows that it's nightfall. This allows the plant to know the length of night, and thus to know the season. This guides its actions, such as flowering. Its "knowledge," to be sure, is limited and easily fooled. Flower growers can flash red light in the middle of night to trick their plants into flowering on time for Mother's Day. Shining red light on a single leaf is enough to do the trick.[9]

Most plants have a blue receptor that regulates their circadian rhythms, such as their daily opening and closing of leaves. This receptor, cryptochrome, is the same receptor that regulates the circadian rhythms of animals, including humans. It differs from another blue receptor, phototropin, that plants deploy in their tips to grow toward the light. Plants can also get "jet lag." If you artificially shift the time of day when they receive blue light, they take a few days to adjust their rhythms, so that their leaves again open and close in synchrony with the light.[10]

Some plants are photoreceptor show-offs. As I mentioned in the preface, *Arabidopsis thaliana*, a small weed that looks like wild mustard, has eleven types of photoreceptors, more than double the number that we employ.[11]

But *A. thaliana* is upstaged by lowly cyanobacteria, which have colonized the earth for at least two billion years—possibly as long as three and one-half billion years—and generated the oxygen in the atmosphere that allowed animals to evolve. Some cyanobacteria employ their entire bodies as lenses to focus light. And at least one, the cyanobacterium *Fremyella diplosiphon*, boasts twenty-seven different photoreceptors, which it harnesses, in ways not well understood, to intelligently harvest light of many colors.[12]

Color perception has deep evolutionary roots. Discriminating colors is a powerful tool employed by millions of species to decode critical messages about fitness. It's no surprise then that colors are firmly wired into our own emotions. However, our understanding of precise associations between colors and emotions is primitive, and the proposed associations between colors and emotions that we listed earlier must be tested by experiments.

For instance, an experiment by Stephen Palmer and Karen Schloss suggests that people prefer colors that they associate with objects they like, such as the blue of fresh water; they dislike colors that they associate with unsavory objects, such as the brown of feces.[13] These associations between colors and objects are forged over eons by evolution, over centuries by culture, and over decades by personal experience. Palmer and Schloss found that the preference for a color depends on the objects it brings to mind, on how close that color is to the color of each such object, and on the emotional response to each object. This result is a promising start.

It is, however, just a start. The human eye can discriminate 10 million colors. Even if we restrict attention to simple patches of uniform color, as in the experiment by Palmer and Schloss, there are many more links between color and emotion to explore. Patches of uniform color are rare in nature. More frequent are combinations of color and texture, called "chromatures," which have a richer structure, can encode more data about fitness, and can trigger more precise reactions.[14]

For instance, in Figure 23 the four green chromatures share, on average, a similar color of green, but their different textures trigger different reactions (this image can be viewed in full color in the Color Insert as Figure G). The green broccoli looks tasty (if you like broccoli), the green strawberry looks inedible, and the green meat looks disgusting. The solid-green square lacks this precision of emotional punch because its texture is trivial. In like manner, the red chromatures share a similar color of red, but because they have different textures they prompt different emotional reactions.

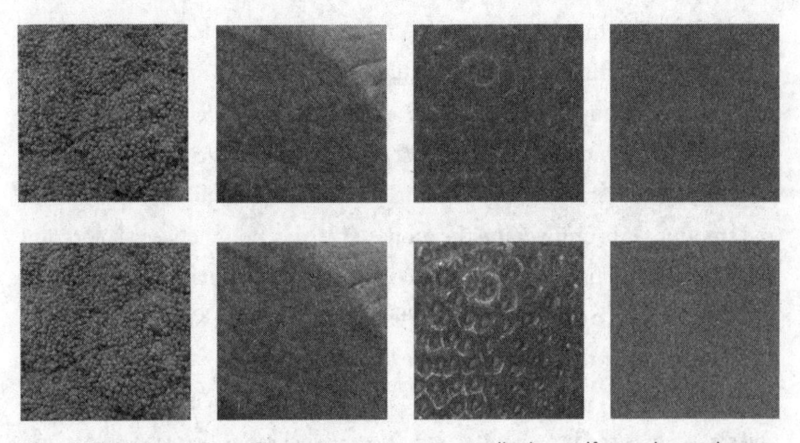

Fig. 23: Eight chromatures. Chromatures are more versatile than uniform color patches at triggering specific emotions. © DONALD HOFFMAN

Although we can discern an impressive 10 million colors, this number pales in comparison to our prowess with chromatures. A square image with just twenty-five pixels can house more chromatures than the visible universe harbors particles, making chromatures a rich channel for messages about fitness.[15] We see hints of this in the chromatures depicted above, which speak eloquently to our emotions with a precision impossible in the patois of uniform colors. The eloquence of chromatures includes nuanced descriptions of shapes, such as the myriad bumps of broccoli and the elegant sweep of a strawberry. These descriptions are carefully crafted calls to action: grasping, squeezing, cradling, pinching, brushing, nudging, grazing, biting, stroking, kissing, and caressing. The eloquence of chromatures extends further, to forecasts of the feedback that can be expected on fingers and lips that answer the call to action: abrasive, bristling, burnished, bulging, chafing, downy, elastic, furry, glassy, hard, icy, jagged, knobbed, limp, moist, numbing, prickly, pocked, ragged, scratchy, slippery, silky, stiff, tingly, unctuous, velvety, woolen, wooden, wet, and yielding.

Chromatures do not pontificate about objective reality—about materials and surfaces of objects that are presumed to exist even if no one looks. Instead, chromatures counsel us how to act, and warn us what to expect, as we forage for fitness. They are a priceless innovation, a compact representation of fitness payoffs, within our species-specific interface. They hide the truth and keep us alive.

For many companies, color is central to branding. We can see this from the golden arches of McDonald's and the red bull's-eye of Target, to the blue bird of Twitter and the green siren of Starbucks. Companies spend fortunes choosing, marketing, and defending their colors. T-Mobile is a wireless phone carrier that spent considerable time and expense branding a specific magenta. AT&T then set up a subsidiary, Aio Wireless, that competed with T-Mobile and featured a plum color in their stores and marketing that was similar to T-Mobile's magenta. When T-Mobile sued Aio for infringement, Aio hired expert witnesses who noted, correctly, that the difference between the plum and the magenta is about twenty times greater than the human threshold for discriminating colors placed side by side. This difference is large enough, they argued, to avoid infringement.

When T-Mobile hired me to reply as an expert, I pointed out that a shopper rarely sees the two colors side by side, but instead must distinguish them from memory. Our ability to distinguish from memory is poor, and the difference between the plum and the magenta is, as it happens, at the limit of our ability. The court agreed with this point, and in February of 2014 issued an injunction against Aio. Federal District Court judge Lee Rosenthal wrote that "T-Mobile has shown a likelihood that potential customers will be confused into thinking that Aio is affiliated or associated with T-Mobile based on the confused association between Aio's use of its plum color and T-Mobile's similar use of its similar magenta color." T-Mobile released a statement saying that the ruling "validates T-Mobile's position that wireless customers identify T-Mobile with magenta and that T-Mobile's use of magenta is protected by trademark law."

As this case demonstrates, color can be prized intellectual property. But a chromature can be far more valuable. Chromatures are more informative than colors, and can be crafted to target specific emotions, or to be congruent with specific products and contexts.

For instance, color psychologists sometimes claim that red encourages appetite. But does it?

Consider the four reds in Figure 24 (this image can be viewed in full color in the Color Insert as Figure H). The first two may whet the appetite, but the last two may trigger disgust. The difference is chromatures.

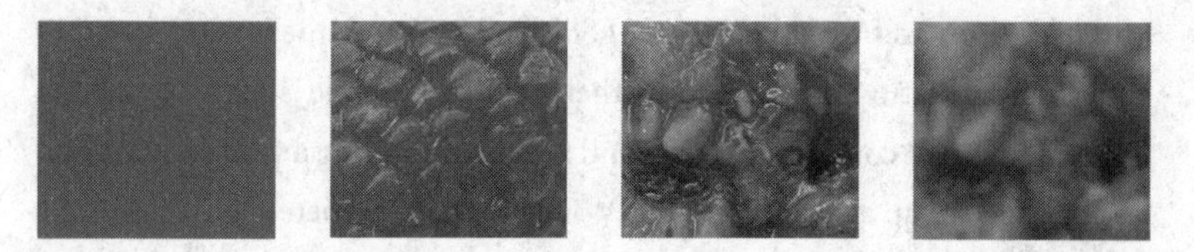

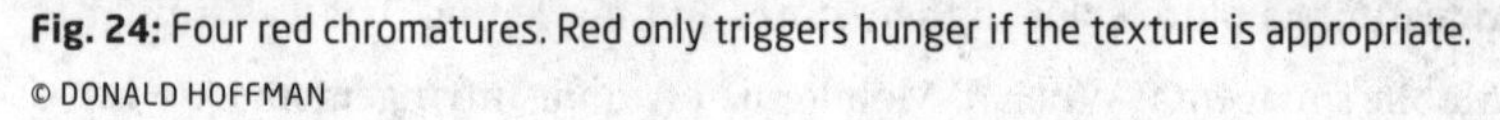

Fig. 24: Four red chromatures. Red only triggers hunger if the texture is appropriate.
© DONALD HOFFMAN

Tomoko Imura and her colleagues have shown that chimpanzees use chromatures to determine the freshness and desirability of fruits and vegetables, such as cabbages, spinach, and strawberries.[16] If you doctor a chromature you can manipulate the emotional reaction of chimps and humans.

Our perceptions are a user interface that evolved to guide our actions and keep us alive long enough to reproduce. Once we grasp this, and free ourselves from the conceptual straitjacket of assuming that we perceive reality as it is, then we can reverse-engineer our interface, understand how it codes information about fitness and guides our actions, and then apply this knowledge to solve practical problems—such as creating chromatures that evoke specific emotions.

It is no small challenge to pull a Houdini and exit our conceptual straitjacket. Thinking about synesthesia, a fusing of senses, can help with this trick. One reason we're sure we see reality, and not just an interface, is that we're sure others see things pretty much the way we do. Suppose I say to you, "That red tomato on the table looks ripe and ready to eat," and you agree. I naturally assume that your perceptions are the same as mine and, indeed, the same as objective reality. Why else would we agree? Surely, it's because we accurately perceive the same reality.

But even if we agree in conversation, we may disagree dramatically in perception. Four percent of humans are synesthetes, who live in perceptual worlds quite foreign to the rest of us.[17]

There are many kinds of synesthesia. In one, each sound of a language triggers a unique experience of color. In his book *Speak, Memory*, Vladimir Nabokov describes his own "fine case of colored hearing": "The long *a* of the English alphabet . . . has for me the tint of weathered wood, but a French *a* evokes polished ebony. . . . I see *q* as browner than *k*, while *s* is not the light blue of *c*, but a curious mixture of azure and mother-of-pearl."[18]

Most of us simply hear the sounds of language, but Nabokov also saw each sound as a specific color, or even a specific chromature, as his descriptions of "polished ebony" and "curious mixture of azure and mother-of-pearl" suggest.

Colors and chromatures appear in a wide variety of synesthesias. They can be triggered by music, printed letters, printed numbers, days of the week, months of the year, emotions, pains, odors, tastes, and even personalities. In "grapheme-color" synesthesia, each symbol for a letter or number is seen as having a color. For instance, A might look red, B might look green, and so on through the entire alphabet.

In gustatory-tactile synesthesia, each taste has an associated shape in three dimensions that can be felt by the hands. The synesthete Michael Watson described his experience of spearmint to the neurologist Richard Cytowic: "I feel a round shape. . . . It's also very cool so it has to be some sort of glass or stone material because of the temperature. What is so wonderful is the absolute smoothness of it . . . the only thing I can explain this feeling as is that it's like a tall, smooth column made of glass."[19]

Watson's experience of other tastes was equally detailed. For instance, angostura bitters: "This definitely has an organic shape. It has the springy consistency of a mushroom . . . it feels like oily leaves on a short vine. I guess the whole thing feels like a scraggly basket of hanging ivy."[20]

Notice what Watson is revealing. He perceives a complex object—a smooth column of glass, a basket of ivy—not as a veridical perception of a mind-independent object, but simply as a useful data structure for representing properties of a taste. Mint is nothing like a column of glass, and angostura bitters is nothing like ivy. This exemplifies the claim of ITP that your perception of a physical object is not a veridical sketch of a preexisting object. It is a data structure that you create as needed to compress critical information about fitness payoffs into an actionable format; once the object has served its purpose, you then garbage-collect its data structure to free up memory so that you can create a new object with your next glance. Contemplating Watson's synesthesia can free our imagination from the chokehold of preexisting objects, from the belief that our object experiences are low-resolution versions of real objects in objective reality.

Music triggers colored shapes in the synesthete Deni Simon, another subject interviewed by Cytowic and Eagleman: "When I listen to music, I see . . . lines moving in color, often metallic with height, width and, most importantly, depth." She explains, "The shapes are not distinct from hearing them—they are part of what hearing is. . . . Each note is like a little gold ball falling."[21]

The artist Carol Steen enjoys several forms of synesthesia. Smells trigger colors. Graphemes, words, sounds, touch, and pain trigger rhapsodies of color, shape, and even movement and location. Her synesthesia streams a torrent of creative visuals from which she ladles inspiration for her paintings and sculptures: "These brilliantly colored and kinetic visions, or photisms . . . are immediate and vivid."[22] Steen describes the bounty of a synesthetic experience: "The shapes were so exquisite, so simple, so pure and so beautiful. . . . I saw a year's worth of sculpture in a few moments."

These synesthetic shapes and colors can be exquisitely detailed. In 1996, Steen sculpted *Cyto*, a maquette in patinaed bronze about eight inches tall, which depicts the complex shapes and chromatures of her synesthetic experience of the grapheme "Cyto." Her experience is not a vague memory or conceptual association, but instead a concrete encounter, a detailed perception. But even her meticulous sculpture omits the dynamic evolution in time of her synesthetic experience, which she describes as shapes that dance.

As these examples illustrate, in many cases a synesthetic experience is not a hazy imagination or weak conceptualization—it's a genuine perception as immediate and compelling as smashing your thumb with a hammer. Notice that Steen is telling us the same important message as Watson: *Cyto* illustrates that Steen sees a precise 3D object, not as a veridical perception of a preexisting object, but simply as a useful data structure for representing, in this case, a particular grapheme.

Synesthetic experiences are consistent over time. A grapheme-color synesthete, for instance, who experiences a specific color for each grapheme of a letter or number, will report the same colors in experiments performed weeks or even years apart. Consistency is used as a "test of genuineness" to discriminate true synesthetes from others who simply invent sensory connections by free association. Some grapheme-color synesthetes report seeing different colors in different parts of single graphemes, while others report seeing the saturation of the colors decrease as the contrast of the graphemes decreases, again suggesting a perceptual rather than conceptual origin.

Synesthesia runs in families, as Francis Galton first noted in the nineteenth century, but the specific associations do not. A parent, for instance, might see the letter A as red whereas their child might see it as blue. Moreover, even the specific senses involved can vary. A parent who sees colors for tastes may have a child who sees colors for graphemes. This suggests that synesthetic associations, although they sometimes involve cultural artifacts such as alphabets and numbers, are not simply taught in families, but are influenced by genetic inheritance.

This is supported by studies of genetic linkage that indicate that synesthesia is influenced by genes on the specific chromosomes known as 2q and 16, and also possibly on 5q, 6p, and 12p.[23] It is too early to reach firm conclusions, but a study of 19,000 subjects suggests that there are five different clusters of synesthesias with different genetic origins—clusters that David Eagleman and his colleagues identify as colored music, colored sequences (such as letters, numbers, months, and days of the week), colors triggered by touch or emotions, spatially displayed sequences, and colors triggered by nonvisual stimuli such as taste.[24]

What are these genes up to? One possibility is that they enhance neural connections between different sensory areas of the brain. In the case of color-grapheme synesthesia, for instance, the cognitive neuroscientists Vilyanur Ramachandran and Edward Hubbard noted that a cortical region in the fusiform gyrus whose activity is correlated with color perception sits next to a region correlated with graphemes.[25] They proposed that synesthetes may be endowed with more neural connections, and thus more crosstalk, between the two regions than nonsynesthetes. This prediction was confirmed by the cognitive neuroscientists Romke Rouw and Steven Scholte with diffusion tensor imaging, which uses magnetic resonance imaging and sophisticated algorithms to estimate connections between regions of the brain in living human subjects.[26] They found that the connections are greater in synesthetes who are "projectors," who see the colors as out in the world, than in synesthetes who are "associators," who see the colors in their "mind's eye." They also found regions in the frontal and parietal lobes that are better connected in synesthetes. No cortical regions were found to be more poorly connected.

Synesthesia is anomalous but not generally pathological. Indeed, synesthetes can enjoy certain cognitive advantages. Some synesthetic associations, for instance, can enhance memory. One grapheme-color synesthete studied by the psychologist Daniel Smilek and his colleagues could recall arrays of numbers better than nonsynesthetes, and her memory improved further when the printed color of each grapheme matched her synesthetic color.[27] Daniel Tammet, an author, speaker, and high-functioning autistic savant, perceives a unique color, shape, texture, and feel for each natural number up to 10,000. Using these synesthetic associations, he memorized and recited more than 20,000 digits of pi—a European record.[28]

Synesthetes beat nonsynesthetes in some perceptual tasks. Michael Banissy found that synesthetes who see synesthetic colors can discriminate between colors better than nonsynesthetes; synesthetes who feel synesthetic touches can discriminate between touches better than nonsynesthetes.[29] Julia Simner and her colleagues studied synesthestes with sequence-space synesthesia—in which sequences such as numbers, letters, days of the week, and months of the year are seen as specific visual forms at specific locations in space—and found that they are better than nonsynesthetes at mentally rotating a 3D object to see if it matches another object.[30]

I started this brief tour of synesthesia with the promise that, at the end, it may free us from a straitjacket—the belief that we see reality as it is. The tour reveals that synesthetes enjoy idiosyncratic perceptions that guide adaptive behavior and are as vivid, complex, and nuanced as our own.

For all we know, Michael Watson's idiosyncratic interface was richer and more adaptive than our own. We do know that it was an aid to Watson in cooking. As Richard Cytowic observed: "He never followed a recipe but liked to create a dish with an 'interesting shape.' Sugar made things taste 'rounder,' while citrus added 'points' to the food."[31] Watson's interface was no less dynamic than ours: "The shape changes with each moment, just as flavor does. . . . French cooking is my favorite precisely because it makes the shapes change in fabulous ways."[32]

We have no grounds for claiming that our interface is veridical and Watson's an illusion. In fact, neither is veridical nor an illusion. Each is an adaptive guide for a critical decision—what shall I put in my mouth? It is an accident of evolution, not a necessity of veridical perception, that Watson's brand of interface is less common. Recall, as we discussed earlier, that some mishap millions of years ago handicapped all vertebrates with an eye of stupid design—our photoreceptors hide behind curtains of neurons and blood vessels that block and scatter light. Cephalopods dodged this mishap and inherited a better model. Perhaps some mishap saddled us with an inferior interface for sensing the quality of foods and, as luck would have it, a mutation gave Michael an upgrade. If, in the future, our survival requires haute cuisine, then natural selection could favor Watson's kind of synesthesia, and future generations might all feel columns of glass when they eat mint.[33]

The point is: we do not have true or ideal perceptions. Instead, we inherit a satisfactory interface with a limited variety of formats—smells, tastes, colors, shapes, sounds, touches, and emotions. Our interface evolved to be fast, cheap, and just newsy enough about fitness to enable us to raise our offspring and pass on our genes. The formats are arbitrary, not the bona fide structures of reality. There are countless formats—other modes of perception—that could serve just as well, or better. We can no more imagine them concretely than we can imagine a specific new color. What is it like to be a bat snatching moths on the wing using sonar? Or to be a moth jamming that sonar in the nick of time?[34] To be a beetle on a bottle, or a moose on a bronze bison, trying to mate? Or a mantis shrimp with twelve kinds of photoreceptors, six for ultraviolet? For these and countless cases, we just don't know. The tinkering of evolution can concoct perceptual interfaces with endless forms most beautiful and most wonderful; the vast majority of these, however, are to us most inconceivable.

Evolution is not finished tinkering with the perceptual interfaces of *Homo sapiens*. The mutations that bless one in twenty-five with some form of synesthesia are surely part of the process, and some of these mutations might catch on; much of the tinkering centers on our perceptions of color. Evolution defies our silly stricture that our perceptions must be veridical. It freely explores endless forms of sensory interfaces, hitting now and then on novel ways to shepherd our endless foraging for fitness.

CHAPTER NINE

Scrutiny

You Get What You Need, in Both Life and Business

"The mind does not pay equal attention to everything it perceives. For it applies itself infinitely more to those things that affect it, that modify it, and that penetrate it, than to those that are present to it but do not affect it."

—NICHOLAS MALEBRANCHE, *THE SEARCH AFTER TRUTH*

Our senses forage for fitness, not truth. They dispatch news about fitness payoffs: how to find them, get them, and keep them.

Despite their focus on fitness, our senses confront a tsunami of information. The eye sports 130 million photoreceptors, which collect billions of bits each second.[1] Fortunately, most of those bits are redundant: the number of photons caught by a receptor differs little, in general, from the number caught by its neighbors. The circuitry of the eye can, with little loss in quality, compress those billions of bits down to millions—just as you may, with little loss in quality, compress a photo. It then streams the millions of bits to the brain through the optic nerve. This stream, though compressed a thousandfold, is no gentle brook. It is a flood, which would overwhelm the visual system if untamed. Taming this flood is the job of visual attention. Billions of bits enter the eye each second, but only forty win the competition for attention.[2]

The initial descent from billions of bits to millions loses almost no information—like a book manuscript edited to omit needless words. But the final plunge to forty loses nearly everything, reducing the book to a blurb. This blurb must be tight and compelling—just the essentials to forage for fitness. This may feel at odds with your own experience of a visual world that seems packed, from corner to corner, with myriad details about colors, textures, and shapes. Surely, it would seem, we see more than just a headline, we see articles, editorials, classifieds—the whole works.

But our experience deceives us. Consider the two images of Dubai in Figure 25. They are identical, except for three major changes. Try to find them. For most of us, it takes a surprisingly long time—a phenomenon known as "change blindness."[3] We hunt in vain, until we happen to stumble upon a difference, whereupon we can't help but see it thereafter. There are many examples online of change blindness, which will entertain you as they demonstrate that it is an important and general aspect of human vision.[4]

Fig. 25: Change blindness. There are three differences between these two images.
© DONALD HOFFMAN

What is going on here? Vision forages for fitness, but the foraging process itself, to be fit, must be lean and only deploy its meager resources with discretion. Countless messages about fitness impinge on the eye, like a thousand emails flooding an inbox. The visual system doesn't waste time and energy reading them all. It treats most of them as spam, and deletes them immediately. It selects a precious few to read and act on. Getting unwanted email on your smartphone is a nuisance and culling it a chore. But with vision the stakes are life and death. One who attends to the frivolous, while missing the vital, will forfeit becoming an ancestor. Natural selection ruthlessly shapes our visual attention to be a nimble forager.

To cut billions of bits to forty, the visual-spam filter is ruthless about deletion. It follows simple and fascinating rules. For those deployed in the trenches of marketing and product design, knowing these rules is essential to success in the ubiquitous battle for the ephemeral attention of consumers. Those who master the rules can direct attention to their products and away from the competition. Those less versed in the rules risk inadvertent altruism.

The opening gambit of the visual filter is its placement of photoreceptors. Unlike the sensor of a digital camera, whose pixels are equally spaced throughout, the retina of the eye deploys more photoreceptors in the center of vision, and ever fewer toward the periphery. Most of us assume that we see the whole field of vision in rich detail. But we're wrong, as Figure 26 demonstrates. If you look at the dot in its center, then you will see that smaller letters in inner rings are as easily discerned as larger letters in outer rings. To be equally legible, the letters in outer rings must be larger, because there the density of your receptors is lower.

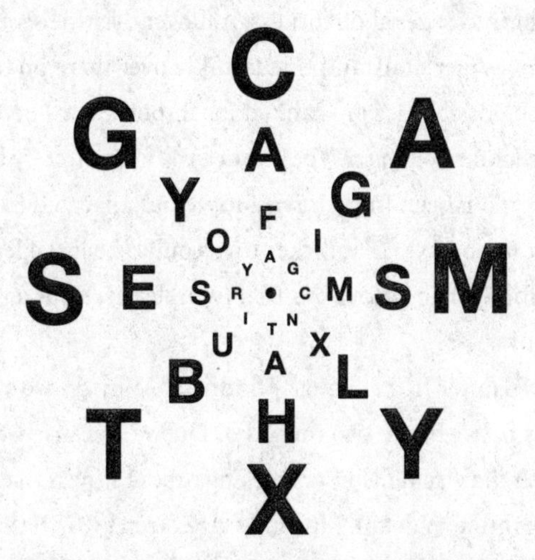

Fig. 26: Visual acuity. If you stare at the middle dot, the big letters are as clear as the smaller. © DONALD HOFFMAN

As you can see from the figure, the density of photoreceptors drops rapidly. Indeed, although our visual field extends two hundred degrees horizontally and one hundred fifty degrees vertically, we enjoy high resolution in only the two degrees that surround the center of gaze. The visible width of your thumb when you see it extended at arm's length is one degree. As I've mentioned earlier, staring at your thumb on your outstretched arm brings home how tiny your window of detail really is: its area is ten thousand times smaller than your field of vision.

Why is it, then, that most of us never notice this limit of vision, and mistakenly believe that we see the whole field of vision in high resolution? The answer lies in the incessant movement of our eyes. They look and jump, look and jump, about three times a second—more when you read, less when you stare. The looks are called fixations and the jumps are known as saccades. Each time you look at something, you view it through a tiny window replete with detail. Normally you don't look and see a blur. So we find it natural to assume that we see everything, at once, in great detail.

The placement of photoreceptors is part of an inspired strategy in the quest for fitness. The wide field of vision, with its low resolution, is used to hunt for possible messages about fitness. A flicker over there on the left might be the twitch of a tiger's tail, and that twinkle over there on the right might be water. These possibilities are ranked for importance—better check for a tiger before checking for water. Then your eyes look directly at each item in order, so that each is seen in high resolution and analyzed in enough detail to decide what to do next. That flicker turns out to be just a leaf in the wind, not a tiger, so forget it and move on. That twinkle turns out to be water. Time to go get a drink.

Why do we suffer from change blindness? Why do we struggle to find the differences between the two images of Dubai? Because we forage for fitness. We search the visual field for a message about fitness that may be worth the effort to examine in detail. Most messages aren't worth this effort. Natural selection has shaped us to ignore them. If we ignore them, then we are unlikely to notice if they change. Change blindness is not a failure to see the

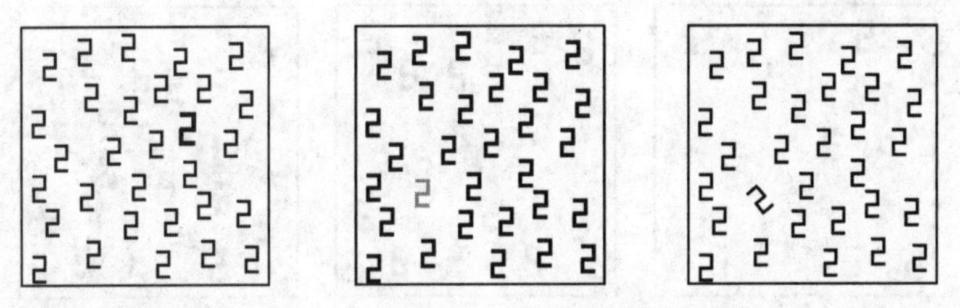

Fig. 27: Pop out. We easily see the large 2 in the left box, the lighter 2 in the middle box, and the tilted 2 in the right box. © DONALD HOFFMAN

true state of objective reality, it's a choice to discard news about fitness that's unlikely to alter our fitness.

For those readers interested in marketing and business, this idea applies to visual advertising. The goal of successful advertising is not merely, and sometimes not even, to present important facts. It is to craft a visual message that rivets the foraging eye of the typical shopper. Consumers face a chaos of competing messages. The trick is to grab their attention. At the simplest level, a message can grab attention by differing from its neighbors in color, size, contrast, or orientation.[5] For instance, in Figure 27, going from left to right, what grabs attention is the larger 2; the 2 of different contrast; the 2 with a different orientation.

In these examples, the item that is different grabs attention quickly even if many items surround it. For instance, in Figure 28 the green 2 "pops out" when there are few distractors, as in the image on the left, but also when there are many distractors, as in the image on the right (this image can be viewed in full color in the Color Insert as Figure I).

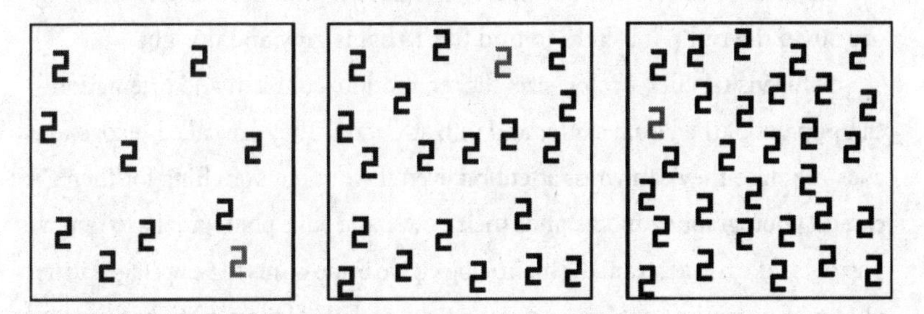

Fig. 28: Color pop out. The green 2 is easily seen even when surrounded by many black 2s. © DONALD HOFFMAN

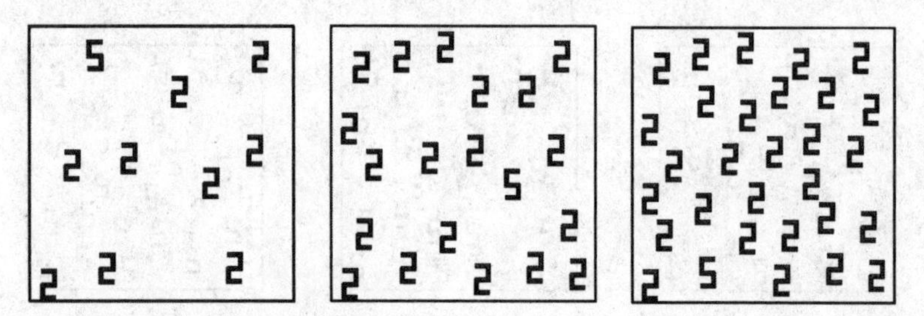

Fig. 29: Difficult search. The 5 in each box does not pop out. One must search for it. © DONALD HOFFMAN

But some differences don't pop out. In Figure 29, the 5 is hard to find, and gets harder with more items around it, as in the image at the right.

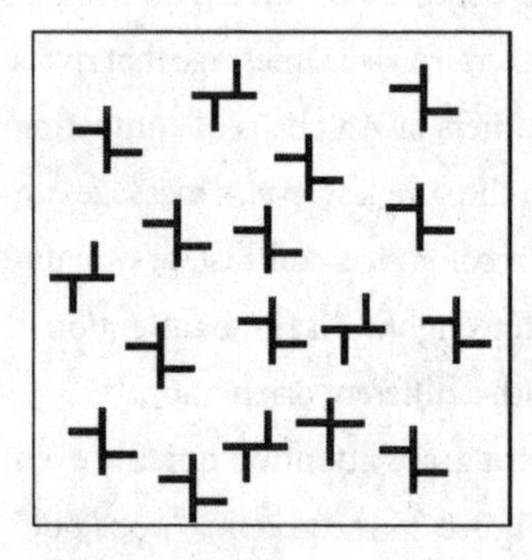

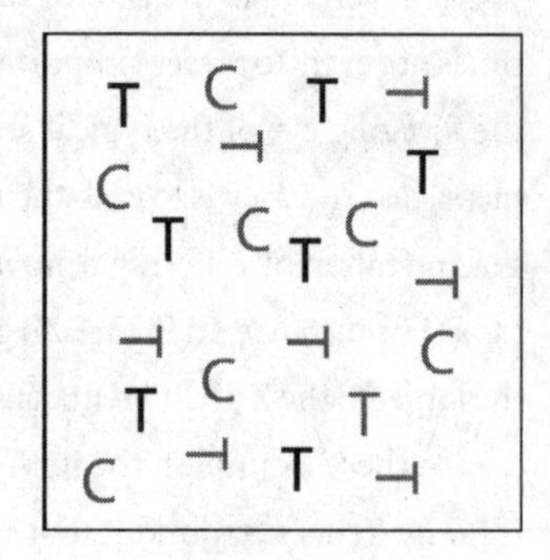

Fig. 30: Difficult search. The cross in the left box and the gray upright T in the right box do not pop out. © DONALD HOFFMAN

Similarly, in Figure 30 on the left, it is hard to find the cross. And in Figure 30 on the right, it is hard to find the T that is gray and upright.

Some visual cues—color, size, flicker, motion, contrast, and orientation—can pop out of the visual clutter and into attention. They are called "exogenous cues" because they can wrest attention even if we're not searching for them. A careful photographer understands their power and edits photographs to remove pop outs that distract from the main subject. No bride wants to be upstaged in her photos by a stray line or high-contrast knickknack that loiters in the background and lures the eye away. The edge of a photograph can itself pop out if it has high

Fig. 31: A store window display. This display makes it difficult to find brand or product information. © DONALD HOFFMAN

contrast. Photographers will sometimes vignette a photograph, gently darkening it near its edges, to remove this distraction and keep the eye on the central subject.

Managing the power of pop out is critical to success in advertising. Every ad, without exception, dictates a foraging strategy for the eye of the viewer. Does your ad send the eye on a goose chase? Or does it guide the eye to glean the facts and emotions you wish to convey?[6] If we think that vision is just a camera that records objective reality, then we misunderstand what really happens when someone views an ad. Think instead of vision, and all of our senses, as foraging instruments evolved by natural selection to hunt for critical information about fitness.

Figure 31 shows a display at the entrance to a sportswear store in an upscale mall. It peppers the eyes with sidetracking cues (this image can be viewed in full color in the Color Insert as Figure J). Most egregious are the bright reflections on the window in the upper left and upper right, and lesser reflections scattered throughout. Their contrast, in brightness and color, lures the eye to dead ends.

When the viewer walks, the reflections slide along the window, and this motion adds to their pointless lure. The cure is reflection-free glass.

But even without reflections, this display echoes with spurious cries from all quarters of a visual jungle. There is a rain forest, two Jackson Pollocks, a wall of non-sequitur orange, stark highlights on bald heads of stiff mannequins and, on the left, hanging by one hand, a dangling modifier—all pointless distractions. There is, if you look closely, a key message: "QUICK DRYING AND VENTED FOR ANY ACTIVITY." Tee shirts on mannequins, meant to be the stars, languish in obscurity for lack of light and contrast.

If vision, like a camera, recorded each detail, then this display might succeed; the data are all there. But vision is no passive camera. It is an impatient hunter for fitness payoffs. It may hazard an unrewarded glance or two at this display, but then give up and move on long before it chances on the key, but hidden, message about drying and venting.

By contrast, the famous ads for iPods expunge all needless pop outs. In these ads, the background splashes a bold, but uniform, color; the foreground sports an ecstatic dancer in black silhouette, devoid of all features, save one: white earbuds sprout white wires that sweep, carefree, down the black silhouette and converge into a white iPod grasped by a gyrating black hand. The emotion is contagious. No words needed, no words used. The message for fitness is clear—iPod equals ecstasy: any questions?

In our visual search for a message deserving attention, we group messages that have common themes, making them easier to attend or discard en masse. For instance, the sixteen dots on the left of Figure 32 can be grouped, based on contrast, into rows, as in the middle, or into columns, as on the right.

Fig. 32: Grouping by brightness contrast. We see horizontal groups in the middle figure and vertical groups in the figure on the right. © DONALD HOFFMAN

They can be grouped by shape, as in Figure 33.

Fig. 33: Grouping by shape. We see horizontal groups on the left and vertical groups on the right. © DONALD HOFFMAN

They can be grouped by size, as in Figure 34.

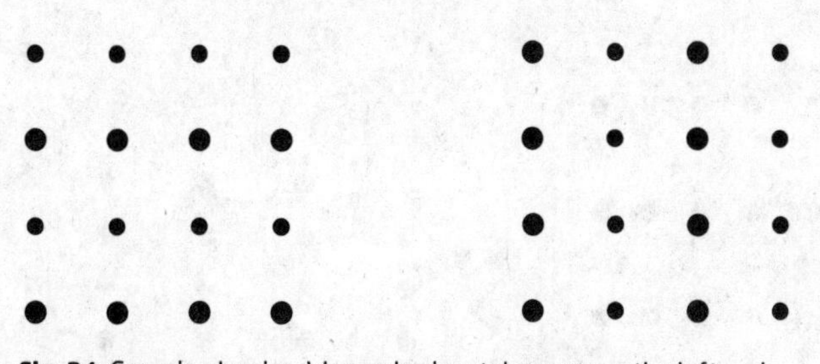

Fig. 34: Grouping by size. We see horizontal groups on the left and vertical groups on the right. © DONALD HOFFMAN

They can be grouped by color, as in Figure 35 (this image can be viewed in full color in the Color Insert as Figure K).

Fig 35: Grouping by color. We see horizontal groups on the left and vertical groups on the right. © DONALD HOFFMAN

They can be grouped by orientation, as in Figure 36.

Fig. 36: Grouping by orientation. We see horizontal groups on the left and vertical groups on the right. © DONALD HOFFMAN

They can be grouped by proximity, as in Figure 37.

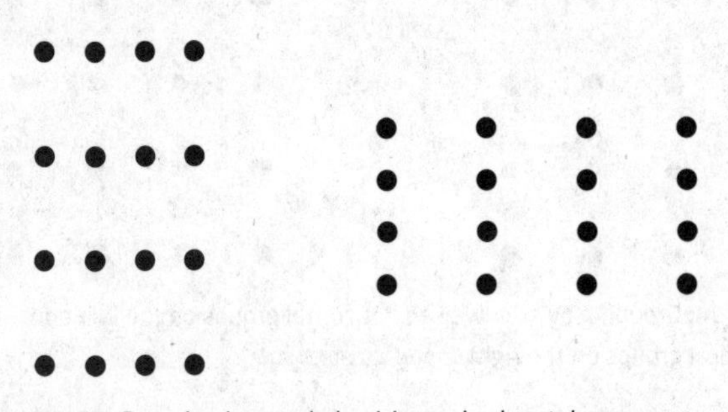

Fig. 37: Grouping by proximity. We see horizontal groups on the left and vertical groups on the right. © DONALD HOFFMAN

This list omits other potent features, such as flicker, motion, and depth.

Competing features can promote competing groups. In Figure 38, on the left, orientation and proximity cooperate to create horizontal groups. But on the right, proximity overrides orientation and dictates vertical groups.

Grouping assists the search for outliers. In Part A of Figure 39, it takes effort to find the maverick line segment. But rearrange the segments to promote grouping, as in Part B of Figure 39, and the outlier pops out. This technique applies to in-store merchandising. A shelf of products can present the shopper with a bewildering mess. But with clever grouping of colors, contrasts, and other features, that shelf can offer happy hunting.

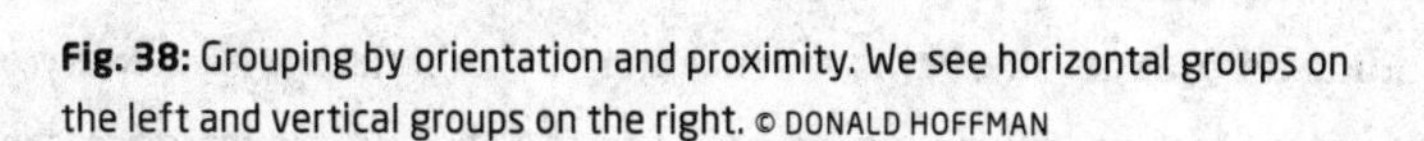

Fig. 38: Grouping by orientation and proximity. We see horizontal groups on the left and vertical groups on the right. © DONALD HOFFMAN

Grouping is a form of data compression. For instance, each line segment in Figure 39 has an orientation, and in Part A of Figure 39, the visual system is forced to describe the orientation of each segment, one at a time. But in Part B of Figure 39, the visual system can make its description much more compact: the eighteen segments on the left are horizontal and the eighteen segments on the right are vertical, except for one at a slant. Grouping lets one description apply to an entire group; no need to repeat the description ad nauseam for each item in the group. This compression helps us find pertinent changes; in Part B of Figure 39, the slanted segment pops out.

Attention is yanked by exogenous cues, but it can be bridled to track endogenous goals. If you search for a lemon, then all things yellow become more salient, aiding your search. Neural activity in area V1 of your brain's occipital cortex correlates with saliency, and with its modification by goals.[7]

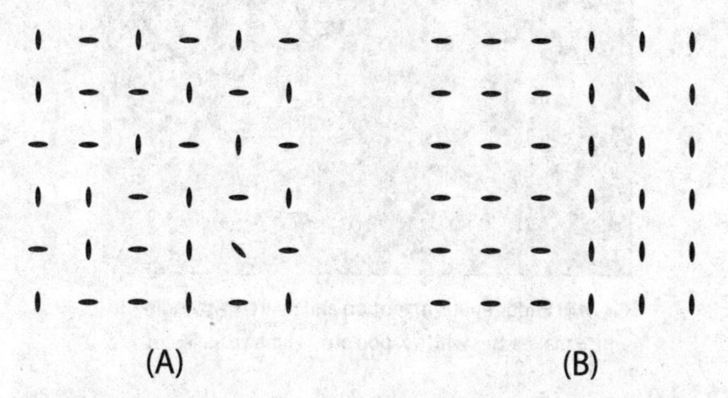

Fig. 39: Grouping and search. It is easier to find the tilted line on the right than on the left. © DONALD HOFFMAN

Nearby neurons signal nearby points in the visual world, so that the whole collection of V1 neurons forms a topographic map of the visual world—a salience map. A neuron actively responding to a feature, such as a color, inhibits nearby neurons if they, too, are responding to that color; this lateral inhibition reduces the salience of those features more common in the field of view, and enhances the salience of the rare. An endogenous goal, such as finding an orange, alters this salience map by enhancing the activity of neurons that respond to features relevant to the goal. If, for instance, you look for black in Figure 40, then a field of black X's occupies your attention. If, instead, you look for white, then a field of white O's enters your attention, and a white X pops out.

If your goal is to check for a tiger hiding in the brush, then your target displays a variety of colors. If you pick the wrong color to enhance in your map of salience, your mistake could end your life. So natural selection has shaped us to enhance colors intelligently. The yellows on the tiger, which

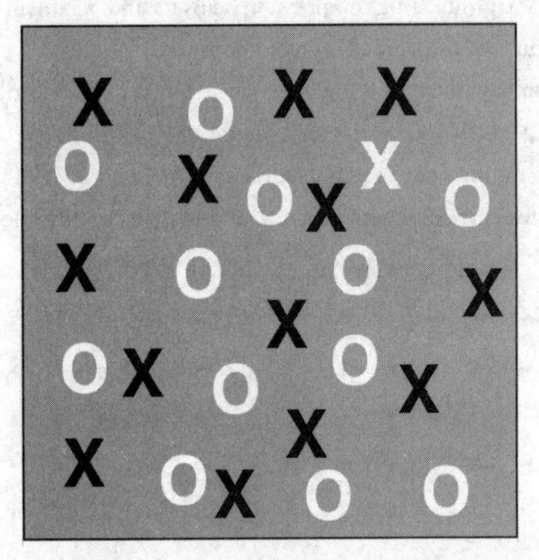

Fig. 40: Endogenous attention and search. Attending to white makes the white X pop out. © DONALD HOFFMAN

match the colors of the brush, are the wrong choice, because enhancing them does little to distinguish tiger from brush. Instead, you astutely enhance the distinctive oranges of the tiger, helping the tiger stripes to pop out visually from the brush, so that the tiger won't pop out viscerally onto your torso.[8]

Enhancing the right features of your target does not, however, guarantee that it will pop out from a scene. You may need to search a bit before your eye nabs your visual quarry—say, predator or prey. If you can search quickly, you are more likely to find your prey in time to put it on your menu, or to find your predators in time to cross yourself off theirs. For this reason, natural selection has shaped your search to be efficient. Your eye looks only to regions rich in distinctive features of your target. And it rarely looks back. If you check a spot and find no target, then your visual system remembers the spot, and doesn't usually send your eye on the fool's errand of returning to that same spot. This handy trick is called "inhibition of return."

It is handy, but not infallible. Suppose you are hungry and searching for a ripe apple. Your visual system duly enhances those regions of your salience map that exhibit the distinctive features of such an apple—say, its red color. Then it picks the spot in your visual field that has the most salience. It directs your eye to look at that spot, to place it in the small window of detailed vision. Then it decodes the fitness message that it finds there. Suppose that the resulting message is *red leaf*. That could be a useful message if, say, you were looking for tinder to start a fire. But you're hungry and want an apple, so *red leaf* doesn't fit the bill. Your visual system diligently triggers its inhibition-of-return trick, so that it won't stupidly revisit that leaf, and then sends the eye to the next spot of interest, the spot with the second-most salience. Suppose there it finds *red rock*. Ah. Not an apple. No need to check there again. Inhibition of return. Everything, so far, is going swimmingly. On to the next spot. Decode the new message. New message: *tiger*. Ah. Not an apple. No need to check there again. Inhibition of return . . .

Whoops! If what you see is not what you seek, then, in most scenarios, inhibition of return is a smart move. But here it could be your last mistake. *Tiger* isn't the message you sought, but it's a message you can't ignore. And not just *tiger*, but any message involving predators or prey. If a hunter-gatherer seeks an apple, and instead spies a hoof or paw, then inhibition of return is the wrong move.

In short, if I see an animal, be it predator or prey, then I should stop my search for an apple or whatever I'm seeking and instead monitor what is mobile. This logic persuaded evolutionary psychologists Joshua New, Leda Cosmides, and John Tooby to propose, in 2007, that we evolved an "animate-monitoring" system. It is designed to detect and monitor any animal in the visual field. The attentional processes that we have discussed so far—based on exogenous cues and endogenous enhancements—rely exclusively on low-level features, such as color, shape, and flicker. The animate-monitoring system, by contrast, is tuned not to low-level features but to a *category* of objects—animals.[9]

New, Cosmides, and Tooby tested their proposal using change-blindness experiments. On each trial, an observer saw a blank screen, then, for a quarter of a second, a photograph of a complex natural scene, then a blank screen, then the same photograph again, but with an important change—an object was deleted. This sequence of frames kept repeating until the observer detected the change. To keep observers honest, one-third of the trials were "catch trials," in which there was no change.

On some trials, the change was to an animate object: a person or an animal. On other trials, the change was to an inanimate object: a plant, a moveable artifact (such as a stapler or a wheelbarrow), a fixed artifact (such as a windmill or a house), or a vehicle (such as a car or a van).

As predicted, observers detected a change to an animate object more quickly than to an inanimate object—on average, one to two seconds more quickly, a significant speedup. One may wonder if the price of more speed is less accuracy. Hasty can mean sloppy. To the contrary, observers missed only one in ten changes to animate objects, compared with one in three to inanimate objects. We're faster and more accurate at detecting animate objects—for good evolutionary reasons.

In modern urban environments, vehicles are more common and dangerous than animals. Nevertheless, observers are faster and more accurate at detecting changes to animals than to vehicles. This is to be expected, if animate-monitoring was wired into us by evolution long before the advent of vehicles. Our eyes forage for fitness today using strategies that our ancestors evolved during the Pleistocene—a geological epoch marked by repeated glaciation, stretching from 2.5 million years ago to just 11,700 years ago.

We can exploit these ancient strategies to design modern marketing. Suppose you sell soap in an orange bottle, and a shopper strolls by, looking instead for a competitor's blue bottle. She glances at your bottle, determines that it's not the color she seeks, flings a dollop of inhibition of return at your shelf of orange bottles, and henceforth ignores your product. That helps her search and hurts your sales.

What to do? How can you disrupt her search for a blue bottle and focus her invaluable attention on your orange bottles? You could trigger her animate-monitoring system. One way would be to stamp, say, a cat or a deer, on your bottles. This could work. But it's far from subtle, and once the competition caught on, they could slap some animal on their bottles, and erase your competitive edge.

To be more subtle, you can dispense with flaunting a beast in toto, and opt instead to reveal some part—an eye, a hand, a paw, a face. A glimpse of an eye is, for purposes of triggering the animate-monitoring system, a glimpse of the beast peering through that eye.[10] Natural selection has made it so: one who attends to a beast only when seen in its entirety risks missing a potential—or becoming an actual—meal. A message that says *eye* also says that there is a creature who owns that eye and warrants your attention.

This advertising strategy—use part of the animal, not the whole—is indeed more muted, but still not subtle enough. The competition will figure it out.

The logic of evolution suggests a better strategy. It takes time to verify that what you see is an eye. If you take too much time on verification, you may fail to act in time to catch a meal, or to avoid becoming one. So natural selection favors shortcuts: anything remotely like an eye wins attention, if only briefly.

The male jewel beetle, you will recall, is lax about what constitutes a significant other. He's just as happy with a glossy bottle as he is with a female beetle. A male moose is tantalized by either a female moose or a bronze bison. A herring-gull chick seeks sustenance from its mother, or from a rectangle of cardboard sporting a red disk. A graylag goose is content to sit on its own eggs, or to try its luck with a volleyball. A male stickleback intent on defending its territory will fight another male, or it will fight a piece of wood with a shape unlike a fish, if it is painted red underneath. Ethologists have a treasure trove of such examples. Natural selection routinely shapes perception to deploy categories that are loose.[11]

This opens a world of possibilities, now largely untapped, for disruptive innovation in marketing and advertising. The eye of the shopper, like that of the beetle and moose, counts on shortcuts and tricks to guide its attention.[12] Those who know its heuristics can lure it at will with well-crafted icons. The trouble, and opportunity, is that little is known of the tricks and shortcuts deployed by human vision to detect animate objects. What simplified icons can still trick the shopper to see, if just for a moment, a face, a hand, an eye, or a butterfly? We don't know. Several years ago, I was strolling down an aisle in a store, and my eyes were suddenly riveted by a bottle of shampoo sporting an annulus that sparkled with iridescence. The exogeneous cue of sparkling, no doubt, grabbed my attention. But I found that I persisted in gazing at that annulus. Perhaps a sparkling annulus says "eye" to the part of vision that triggers the monitoring of animals? What other simplified icons for eyes might trigger such monitoring? And not just icons for eyes, but for the variety of bodily parts of humans and other animals? To answer these questions we must reverse-engineer, with careful experiments, the heuristics that natural selection has wired into human vision.

I have understated the real potential here. The jewel beetle doesn't just like a beer bottle as much as a female; he likes it far more. The herring-gull chick doesn't just like a cardboard-cum-disk as much as its mom; as the disk gets bigger he likes it far more. A stickleback doesn't just fight a red-bellied blob as much as another male; as the faux belly gets bigger he will ignore a real male to fight the harmless blob. A male *Homo sapiens* doesn't just like a female with breast implants as much as a female au naturel; if the implants impart an upper convexity not seen in nature, he likes it far more.[13] A caricature of a face isn't just identified as well as a photograph, it is identified more quickly.[14]

These are examples of "supernormal stimuli."[15] Evolution shapes the perceptions of an organism to track fitness—not truth—as cheaply as possible given the demands of its niche. Supernormal stimuli hint at the resulting codes for fitness. In its niche, a herring-gull chick can succeed with a simple code: a larger red disk means a better chance for food.

The implications for marketing are clear. A simple icon, crafted to exploit the visual codes wired by natural selection into the visual systems of consumers, can grab attention with supernormal power. Such an icon can be subtle and thus difficult for a competitor to reverse-engineer, and yet highly effective. For icons used in branding, emotional import is also critical. The goal is not just to grab attention, but to grab the right kind. This typically requires an icon that associates with the brand a specific, positive feeling—say, prestigious and wealthy, or rugged and healthy. An icon that brandishes fangs will grab attention, but—apart from ads for vampire movies and Halloween costumes—attention of the wrong kind. A well-crafted icon can exaggerate, judiciously, visual features that draw attention and trigger a desired feeling.

Suppose, for instance, you want an icon of an eye that grabs attention and feels attractive. Recall, from chapter two, that a female eye looks more attractive if it features a large iris, a dilated pupil, a bluish sclera, conspicuous highlights, and a prominent limbal ring. There are surely other critical features of an attractive eye not yet discovered. The challenge for a marketing team is to create an icon—perhaps a stylized eye, or something more abstract—that captures such features with supernormal effect. At present, given the limits of our scientific knowledge, this challenge may best be achieved through the intuitions and talent of a graphic designer. But a corporation that conducts experiments, guided by evolutionary theory, to learn how to hack the visual code of *Homo sapiens* for the attractiveness of eyes, could exploit its knowledge to create icons that manipulate this code to powerful effect.

This is just one example in a vast and largely unexplored territory. One-third of the brain's cortical activity is, as we have discussed, correlated with visual perception. If you include the other senses, there's lots of sensory coding to explore and hack. Some of it, perhaps most of it, is spaghetti code, as inelegant as the unintelligent design of our eye, with its photoreceptors stupidly cloaked behind neurons and blood vessels. Our perceptions are a species-specific user interface, not a window on truth, and its underlying code is a sea of kludges, punctuated by islands of inadvertent brilliance. Vision does not approximate an ideal observer who recovers objective truths. It is an interface kludged together on the cheap. It tells us just enough about fitness to keep us alive in our niche long enough to raise kids. Understanding this, and letting it guide our choice of experiments, is a promising direction for perceptual science, marketing, and product design.[16]

Our interface is wired to detect and monitor predators and prey. The logic of selection that installed this wiring is, we have seen, clear and compelling—those with the wiring are more likely to enjoy lunch than to be lunch. Meat, however, was not alone on the menu of *Homo sapiens.* We are omnivores, not just carnivores, and our ancestors have long eaten fruits and vegetables. Has natural selection wired us to detect fruits and vegetables and, since they are immobile, to remember where they reside?

The evidence for preferential detection of fruits and vegetables is, at present, equivocal. The experiments by New, Cosmides, and Tooby, which found quick detection for animate objects, found otherwise for plants. However, the plants they tested were trees, shrubs, and a pineapple. No experiment, to date, has studied whether we are specially tuned to detect fruits and vegetables.

The recent evolution of trichromatic vision in primates, which allows finer discrimination between reds and greens, may have been selected in part to aid the detection of ripe fruit against green foliage. This hypothesis, though intriguing, is for now controversial.[17]

However, Joshua New and his colleagues found, in an experiment that took place at a farmer's market, that we remember well the locations of foods, and we remember better the location of a food that is higher in calories (even if that food is not well liked); moreover, women remember better than men.[18] This makes sense. Memory, like perception, evolved in service of fitness. Our memories are no more a veridical report of the past than our perceptions are of the present. Memory and perception don't deal in objective truths. Both deal in fitness, the only coin of the evolutionary realm. It is no surprise that fruits and vegetables that offer more fitness receive more memory.

This suggests that an icon of food can enhance our memory for a product just as an icon of an animal can enhance our attention to it. Care must be taken, of course, to fashion an icon that succeeds at hacking into our visual code and masquerading as a high-fitness food. Get it wrong, and an icon can brand a product as unpalatable and unmemorable.[19] Get it right, and an icon can go supernormal. Add a chromature of a high-end food, such as a honeycomb, and it may make the memory much stronger.

Let's recap. Our eyes are reporters on the fitness beat, searching for a scoop, looking for intelligence about fitness that is worth decoding. A message, once decoded, typically appears in a standard format. We see the decoded message as an object in space, whose category, shape, location, and orientation inform us how to act to glean the fitness points we need. We gumshoe for fitness on the cheap, attending to just a fraction of the leads on offer. Exogenous cues can grab our attention: depth, flicker, and movement; contrasts in size, color, brightness, or orientation. Endogenous goals can alter the salience of exogenous cues. Looking for a pear makes its distinctive green more salient. We constantly scan for anything animate. We may also scan for high-calorie foods. This repertoire of strategies in our search for fitness payoffs makes the process of searching itself more fit.

But we have another technique in our repertoire: scripted attention. Its impact is best described by example. I was asked by a major jeans company to evaluate their new print ad. It prominently displayed a buff man wearing jeans and a winning smile. This is a good move because it triggers, in shoppers, the module of attention that monitors people and animals, and associates with the brand the positive attributes of robust health and upbeat mood. The ad splashed the company logo in bright color and high contrast, a good way to grab attention with exogenous cues. But the ad, to its detriment, misdirected the attention of the shopper, because it missed the role of scripted attention.

Here's how. We are a social species. When you forage for fitness, you note where others forage. Where they attend, you also attend. After all, what grabs the attention of another person may warrant your attention as well. Perhaps they see vital information about fitness that you missed: a stalking lioness, a delicious morsel, a helpful friend, an implacable foe. You infer—from the direction of their body, face, and eyes—where they attend, and you shift your attention to match theirs.

In the jeans ad, the body, face, and eyes of the model all aimed one direction—away from the logo, and into empty space. The model turned his back on his own ad. His body, from head to toe, told the shopper a clear message: forget this product—there's something of greater interest over there, on the left. If, by chance, on the left there was an ad for the jeans of a competitor, then the model would unwittingly tell shoppers that the competitor's jeans deserve more attention than his own. This is not the best use of marketing dollars.

Fortunately, this was easy to fix. I swapped the two sides of the ad, so that the model directed attention where the jeans company wanted it—on their logo. This is an example of scripted attention: we use our knowledge of our current context to constrain how we forage for fitness, allowing us to forage with greater speed and precision. In the context of viewing a person, our script leads us to attend where the person's face and body appear to be focused.

We deploy other scripts for attention. In a store, you don't search for products on ceilings or floors; you just attend to shelves. In your bathroom, you know where to look for soap and razor. If you're driving in the US, then you glance left before turning right; in the UK, you do the opposite. If you fly from the US to the UK and rent a car, good luck—your scripts, I can attest, dispatch your attention to random places, risking mayhem. A script for attention that buoys fitness in one context can scuttle it in another. Natural selection shaped in us the capacity to learn new scripts; as the environment changes we can alter our scripts.

Our script for people directs us to follow their gaze. But it does more. It directs us to look at hands. What is that hand up to? Where is it pointing? What is it holding? A weapon? Food? The hand of another person can, in an instant, alter your fitness for better or worse. Attending to hands is itself a fit strategy. In the jeans ad that I evaluated, the hands of the model did nothing to promote the product. They just dangled. If, instead, a hand is made to hold a product, or gesture toward a logo, then that hand can assist in directing attention.

Standard accounts of attention assume that objective reality consists of cats, cars, and other physical objects in space and time, and that attention directs us to look at, and to accurately perceive, these preexisting objects. This assumption is false. Cats and cars are messages about fitness in the sensory interface of *Homo sapiens.* When I look from cat to car, I don't switch attention from a preexisting cat to a preexisting car. Instead, I decode one fitness missive and get the message *cat*, then I decode a second missive and get the message *car.* I create and then destroy cat and car and other objects as needed, in my endless foraging for fitness.

Fitness functions are complex, depending on the organism, its state, its action, and the state of the objective world (whatever that world may be). Some aspects of fitness are stable. That is why I can see my cat Tulip, look away, then look back and see her again. I see the same Tulip because I decode the same missive about fitness. Some aspects of fitness are transitory. If I take a step to the side and then look again at Tulip, she looks a bit different, a bit rotated. If I eat two hamburgers, a third hamburger doesn't appeal to me quite as much as the first two. These variations in my perceptions of cat and burger reflect variations in the fitness that these objects encode.

I love my cat and enjoy my car. But I don't believe they exist if unperceived. Something exists. Whatever that something is, it triggers my senses to acquire a coded message about fitness in an idiom of cats, cars, and burgers—the parlance of my interface. That vernacular is simply inappropriate to describe objective reality.

I love the sun and don't want to part with my neurons. But I don't believe the sun existed before there were creatures to perceive it, or that my neurons exist if unperceived. Stars and neurons are just icons in the spacetime desktop of my perceptual interface.

If our senses were shaped by natural selection then our perceptions do not portray true properties of objective reality, any more than the magnifying-glass icon in my photo-editing app portrays the true shape and location of a real magnifying glass inside my computer. When I click on that icon my photo enlarges. If I ponder why it enlarges, I may conclude that the icon is the cause. I would be wrong. My mistake is a harmless and even useful fiction, as long as I just edit photos. But if I want to build my own app, then this fiction is no longer harmless. I need to understand a deeper level of cause and effect within the computer that is hidden by its interface. Similarly, for most research and medical applications it is a harmless and even useful fiction to think that neurons have causal powers—that neural activity causes my thoughts, actions, and other neural activity. But if I want to understand the fundamental relationship between neural activity and conscious experiences, then this fiction is no longer harmless. I must understand a deeper level of cause and effect that is hidden by the spacetime format of my sensory interface.

The reason that my perceptions can't show me the truth, can't show me the sun-in-itself, is that the sun-in-itself is shrouded by a cloud of fitness payoffs. This cloud determines my fate and the kismet of my genes. Evolution has steadfastly directed my perceptions to the cloud of fitness payoffs, not to the sun-in-itself. The sun-in-itself affects the cloud and, in consequence, my perceptual experience as of the sun, but my perceptual experience as of the sun does not describe the sun-in-itself. A computer file affects its icon on the desktop, but its icon does not describe the file.

Our perceptions of objects in spacetime are not objective reality—the thing-in-itself—nor do they describe it. Does this mean that objective reality is forever beyond the reach of science? Not necessarily.

CHAPTER TEN

Community

The Network of Conscious Agents

"Silence is the language of god, all else is poor translation."

—JALALUDDIN RUMI

"What can be said at all can be said clearly; and whereof one cannot speak thereof one must be silent."

—LUDWIG WITTGENSTEIN, *TRACTATUS LOGICO-PHILOSOPHICUS*

The delight of mystery, which we sometimes fetch from the netherworld of a black hole or a parallel universe, can be enjoyed, here and now, in your very chair. No mystery of science offers more intrigue, or greater perplexity, than the provenance of quotidian experiences—the taste of black coffee, the sound of a sneeze, the feel of your frame pressed into your chair. How does your brain serve up this magic? With what wave of a wand does three pounds of meat beget a conscious mind? That this remains a mystery is not, it would seem, due to a dearth of data: scientific journals are packed with scan upon sundry scan of a brain caught in the magician's act. It's rather that this cagey magician, despite unblinking scrutiny of its act, has never revealed any secrets. For Thomas Huxley in 1869, its legerdemain could be fathomed no better than the magic of Aladdin's lamp. For us today, despite the breakthroughs of neuroscience, it remains just as surely unfathomable.

Why are we stumped? We can blame that basic tool of the conjurer's trade: distraction. We have been lured, with potent miscues, to look over *here*—at the brain (or at the brain together with the body interacting with the environment). We have been misled to believe that the brain, or the embodied brain, somehow serves up the magic of consciousness. We have, in short, been duped.

For much of this book, I've sketched out how this has happened. Evolution shaped our perceptions to hide the truth and to guide adaptive behavior. It endowed us with an interface, consisting of objects in spacetime. It let us reason, with frequent success, about cause and effect within that interface. If I hit that cue ball just so, causing it to graze the eight ball over there, then I can pocket the eight ball and a chunk of cash. If I challenge that grizzly bear for the honey in that hive, the odds are that I will forfeit the honey and my life. Our grasp of cause and effect can dictate, in contexts both complex and crucial, our payoffs in fitness: a mate or a jilt, a meal or a miss, life or death. We do, and should, take it seriously. But it is a fiction—albeit a lifesaving fiction. Grasping virtual cause and effect in our interface grants us no more insight into the intrinsic operations of objective reality than grasping virtual cause and effect in a video game—fire this machine gun to obliterate that chopper; brandish this shield to deflect that blow; turn this wheel to steer this truck—grants a video virtuoso insight into the intrinsic operations of the transistors and machine code of her computer.

Physicists realize that spacetime is doomed, as well as its objects.[1] For principled reasons, Einstein's spacetime cannot be foundational in physics. A new theory is required, in which spacetime, objects, their properties, and their fiction of cause and effect, sprout from a more primordial ground.

For most science and technology, this fictional cause and effect is handy—it helps us understand and exploit our interface. But if we try to understand our own conscious experiences, then this fiction gets in the way. Its lure, wired by evolution into even the best and brightest minds, poses the single greatest impediment to our progress. This fiction is built into each theory of consciousness that assumes, in accord with the Astonishing Hypothesis, that consciousness arises somehow from packs of neurons. This fiction is at the core of a proposal by Roger Penrose and Stuart Hameroff that conscious experience arises from an orchestrated collapse of certain quantum states in neural microtubules.[2] It is at the core of a proposal by Giulio Tononi and Christof Koch that each conscious experience is identical to some causal structure, neural or otherwise, that integrates information.[3] None of these proposals has offered a precise account for a single conscious experience. Precisely which orchestrated collapse creates, say, the taste of ginger? Precisely which causal architecture for integrating information is the smell of pine? No answer has been offered and none ever will: these proposals set themselves an impossible task by assuming that objects in spacetime exist when not observed and have causal powers. This assumption works admirably within the interface. It utterly fails to transcend the interface: it cannot explain how conscious experiences might arise from physical systems such as embodied brains.

If no theory that starts with objects in spacetime can account for our conscious experiences, then where shall we begin? What new foundation might allow us to integrate the volumes of hard-earned data on mind, matter, and their correlations, into a rigorous theory? We can rephrase this question with a diagram we first encountered in chapter 7 (Figure 41). Suppose that I am an agent—a conscious agent—who perceives, decides, and acts. Suppose

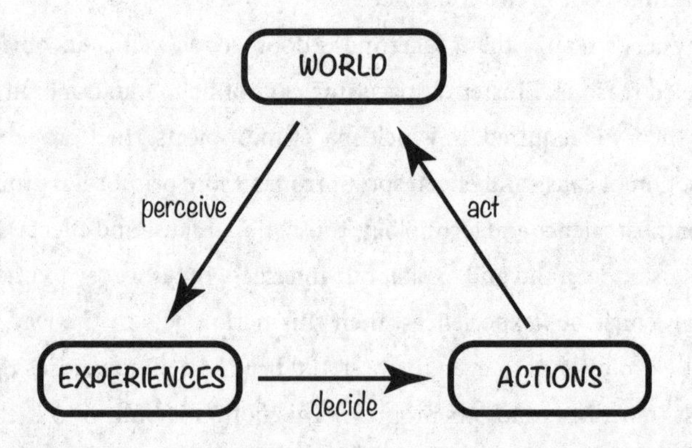

Fig. 41: The "perceive-decide-act" (PDA) loop. © DONALD HOFFMAN

that my experiences of objects in spacetime are just an interface that guides my actions in an objective world—a world that does not consist of objects in spacetime. Then the question becomes: What is that world? What shall we place in that box labeled WORLD?

Now this form of the question itself makes assumptions that may prove false. Perhaps, for instance, I'm just wrong to believe that I enjoy conscious experiences—that I experience the taste of mint tea and the smell of oatmeal cookies, and that I experience myself drinking that tea and eating those cookies. Perhaps there are no such experiences and I am deluded. The issue here is not whether I am infallible in my beliefs about my conscious experiences; the field of psychophysics provides clear evidence that no one is infallible. The issue is that I may be wrong to believe that I have any experience at all.

I cannot rule out this possibility. However, if I am wrong to believe that I have conscious experiences then, it would seem, I am wrong to believe anything. I should just eat, drink, and be merry, and grant that these pleasures themselves are but a delusion.

Let's agree to put aside this possibility for the moment. Let's grant, provisionally, that we have conscious experiences, that we are fallible and inconsistent in our beliefs about them, and that their nature and properties are legitimate subjects of scientific study. Let's also grant that our experiences, some of which we are consciously aware of and many of which we are not, inform our decisions and actions; again, taking these as ideas to be refined and revised by scientific study. Let us grant, in short, that we are *conscious agents* that perceive, decide, and act. The notion of a conscious agent is based on intuitions that are widely shared. It must, however, be made precise and then endure the rough and tumble of science.[4]

Then the question remains: What is the objective world?

Perhaps our world is a computer simulation and we are just avatars that haunt it—as in movies such as *The Matrix* or *The Thirteenth Floor*, and games such as *The Sims*. Perhaps some geek, in another world, gets her kicks creating and controlling us and our world. That geek and her world might in turn be the digital plaything of a geek in a lower-level world. This might continue for multiple levels, until we reach some base level where the first simulation runs. Perhaps that level was conceived by a single edgy artist, or arose as a joint endeavor of a brilliant civilization beyond our imagination, or started as a scientific experiment to test whether new rules of physics could spark fascinating life forms whose creativity and pleasure was worth the pain they suffered.

This possibility is not dismissed by some serious thinkers, such as philosophers Nick Bostrom and David Chalmers, as well as tech entrepreneur Elon Musk, and it has interesting points in its favor. Spacetime, for instance, may be pixelated much like a computer screen; the three dimensions of space are a holographic inflation much like the virtual worlds of video games.

Could conscious experiences bubble out of a computer simulation? Some scientists and philosophers think so, but no scientific theory can explain how. Some suggest that each *specific* conscious experience—such as the taste of coffee I am savoring right now—is a specific computer program. But no such program has been found, and no one has any idea what principle could tie a program to an experience. For now, this proposal is a hand wave, not a scientific theory.

Others suggest that each *kind* of conscious experience—such as the kind of taste I have whenever I drink coffee—is a class of programs. But again, no such class of programs has been found, and no one has any idea what principle could tie a class of programs to a kind of experience. In short, we have no idea how simulations might conjure up conscious experiences. Simulations run afoul of the hard problem of consciousness: if we assume that the world is a simulation, then the genesis of conscious experiences remains a mystery.

It is, as we have seen, an empirical fact that specific conscious experiences are tightly correlated with specific patterns of activity in neural circuits. But no scientific theory that starts with neural circuitry has been able to explain the origin of consciousness. Steven Pinker suggests that we may have to live with this: "The last dollop in the theory—that it subjectively feels like something to be such circuitry—may have to be stipulated as a fact about reality where explanation stops."[5]

Pinker may be right: in our quest to understand the origin of subjective experience, if we start with circuitry then explanation stops. But could some other proposal fare better?

When facing a problem like this, scientists often heed the counsel of a fourteenth-century friar, William of Ockham: choose the simplest proposal that explains the data. This nugget, known as Occam's Razor, is not a dictate of logic like *modus tollens*.[6] It may on occasion lead one astray. At a meeting of the Helmholtz Club, Francis Crick spotted such an occasion and remarked, "Many men have slit their throats with Occam's Razor."

Yet Occam's Razor rightly enjoys stellar proponents. Einstein endorsed it in 1934: "It can scarcely be denied that the supreme goal of all theory is to make the irreducible basic elements as simple and as few as possible without having to surrender the adequate representation of a single datum of experience."[7] The philosopher Bertrand Russell, in 1924, also gave it the nod: "Whenever possible, substitute constructions out of known entities for inferences to unknown entities."[8]

Occam's Razor, applied to the science of consciousness, counsels a monism over an amphibious dualism, a theory based on entities of one kind rather than two. In accord with this advice, most attempts at a scientific theory of consciousness embrace physicalism. The basic constituents of objective reality are taken to be spacetime and its unconscious contents—particles, such as quarks and electrons, and fields, such as gravity and electromagnetism. Consciousness must somehow emerge from, or be caused by, or be identical to, these unconscious entities. Physicalists seek a theory that makes good on the Astonishing Hypothesis that conscious experiences can be generated by packs of neurons, which are themselves cooked up from unconscious ingredients.

As we have discussed, all attempts at a physicalist theory of consciousness have failed. They have produced no scientific theory and no plausible idea of how to build one. In each attempt so far, at just the moment when consciousness pops out of unconscious ingredients, a miracle occurs, and a metaphorical rabbit pops out of a hat. The failure, I think, is principled: you simply cannot cook up consciousness from unconscious ingredients.

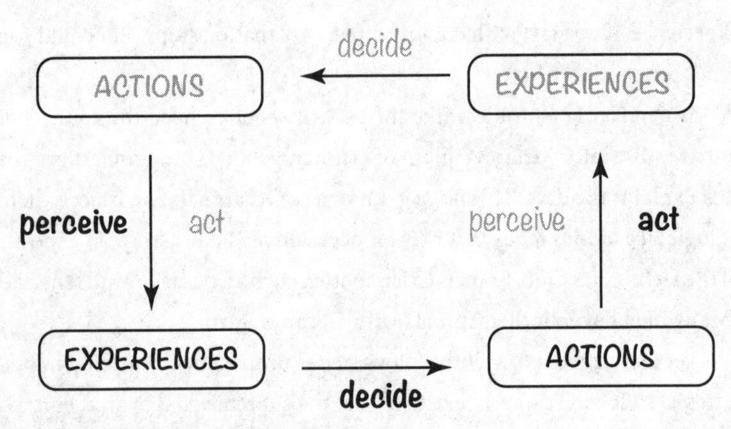

Fig. 42: Two interacting agents. © DONALD HOFFMAN

Physicalism is not the only available monism. If we grant that there are conscious experiences, and that there are conscious agents that enjoy and act on experiences, then we can try to construct a scientific theory of consciousness that posits that conscious agents—not objects in spacetime—are fundamental, and that the world consists entirely of conscious agents.[9]

Consider, for instance, a toy universe with just two conscious agents. Then the external "World" for each agent is the other agent. We end up with two conscious agents that interact. This is illustrated in Figure 42, with one agent in bold type, and the other in light type. How one agent acts will influence how the other perceives; thus, a single arrow is labeled as both *act* and *perceive*.

We can consider universes that are more complex, with networks of three, four, or even an infinity of agents. The way one agent in a network perceives depends on the way that some other agents act. I call this monism *conscious realism*. Conscious realism and ITP are independent hypotheses; one may claim, for instance, that the reality behind our perceptual interface is not fundamentally conscious.

To turn conscious realism into a science, we need a mathematical theory of conscious experiences, conscious agents, their networks, and their dynamics.[10] We must show how conscious agents generate spacetime, objects, physical dynamics, and evolutionary dynamics.[11] We must get back quantum theory and general relativity, and generalizations of these theories that are mathematically precise.

"But," you might say, "anyone who desiccates consciousness into mathematics has, we can safely assume, lost touch with the richness of their own consciousness and vanished into their own pointy head."

Not so. A science of consciousness no more requires divorce from living consciousness than meteorology requires naiveté about thunderstorms, or epidemiology requires disregard for human affliction, or the science of evolutionary games requires virginity. To the contrary, it is fascination with the living subject that inspires a quest for rigor and deeper insight.

"But the proper ontology for science is physicalism. An ontology in which consciousness is fundamental is mere quackery. To reject physicalism, and embrace conscious realism, is to embrace pseudoscience."

Many scientists do, in fact, endorse physicalism. Given that it has, time and again, proven of value in the progress of science and technology, one can hardly fault a scientist who looks askance at other ontologies, such as conscious realism.

Science, however, presumes no ontology. Ontologies are theories, and science—a method for evolving and testing theories—grants to no theory a special dispensation. Each theory, like each species, must compete to endure. A theory that today boasts a long reign may tomorrow, like so many erstwhile species, suffer a sudden extinction.

A certain physicalism that starts with spacetime and unconscious objects has enjoyed a long reign and, because *Homo sapiens* perceives fitness in the argot of objects in spacetime, a prima facie plausibility. But this physicalism appears unfit in some new territories of science, such as quantum gravity and the relation of biology to consciousness. The surprising insight of the FBT Theorem—that an organism that sees objective reality cannot dominate an organism of equal complexity that instead sees fitness—clashes with physicalism and warns of its demise.

"But what about conscious realism? Surely the plausibility of physicalism is surpassed only by the implausibility of conscious realism. Are we really to believe that an electron, which surely feels nothing, is itself conscious or, more outrageous still, a conscious *agent*?"

This objection misinterprets conscious realism, which denies that physical objects exist when unperceived, and denies that they are conscious when perceived; physical objects are our conscious experiences, but they are not themselves conscious. The proper target of this objection is panpsychism, which claims that some physical objects also have consciousness. An electron, for instance, has unconscious properties such as position and spin, but may also have consciousness; a rock, however, might not be conscious, even if it consists of particles that are each conscious. Panpsychism appears unable to avoid dualism.[12] Brilliant thinkers have advocated panpsychism, which underscores the obstinacy of the hard problem of consciousness and the quandary of those trying to solve it.[13]

Conscious realism is not panpsychism. The claim of conscious realism is better understood by looking in a mirror. There you see the familiar—your eyes, hair, skin, and teeth. What you don't see is infinitely richer, and equally familiar—the world of your conscious experiences. It includes your dreams, fears, aspirations, love of music and sports, feelings of joy and grief, and the gentle pressure and warmth in your lips. The face you see in the mirror is a 3D icon, but you know firsthand that behind it is the vibrant world of your conscious experiences that transcends three dimensions. A person's face is a small portal into their rich world of conscious experiences. The curve of lips and squint of eyes that form a smile no more capture the experience of real joy than the letters *j-o-y*. We can, despite this poverty of translation, see a friend smile and share their joy—because we are insiders, we know firsthand what transpires behind the scene when a face fashions a genuine smile. This same advantage of the insider lets us see a frown and feel disgust, see raised brows and feel surprise, and so on, with more than twenty kinds of emotions.[14]

We can convey an experience by a mere expression. This is data compression of impressive proportions. How much information is wrapped up in an experience, say, of love? It's hard to say. Our species has explored love through countless songs and poems and, apparently, failed to fathom its depths: each new generation feels compelled to explore further, to forge ahead with new lyrics and tunes. And yet, despite its unplumbed complexity, love is conveyed with a glance. This economy of expression is possible because my universe of experience, and my perceptual interface, overlaps yours.

There are, of course, differences. The visual experiences of the colorblind differ from the rich world of colors that most of us relish. The emotional experiences of a sociopath differ from ours in a way perhaps inconceivable to us, even in our darkest moments. But often the overlap is substantial, and grants us genuine, if but partial, access to the conscious world of another person, a world that would otherwise lie hidden—behind an icon of their body in our interface.

When we shift our gaze from humans to a bonobo or a chimpanzee, we find that the icon of each tells us far less about the conscious world that hides behind it. We share with these primates 99 percent of our DNA, but far less, it would seem, of our conscious worlds. It took the brilliance and persistence of Jane Goodall to look beyond the icon of a chimp and glimpse inside its conscious world.[15]

But as we shift our gaze again, from a chimp to a cat, then to a mouse, an ant, a bacterium, virus, rock, molecule, atom, and quark, each successive icon that appears in our interface tells us less and less about the efflorescence of consciousness behind the icon—again, "behind" in the same sense that a file lies "behind" its desktop icon. With an ant, our icon reveals so little that even Goodall could not, we suspect, probe its conscious world. With a bacterium, the poverty of our icon makes us suspect that there is, in fact, no such conscious world. With rocks, molecules, atoms, and quarks, our suspicion turns to near certainty. It is no wonder that we find physicalism, with its roots in an unconscious ground, so plausible.

We have been taken in. We have mistaken the limits of our interface for an insight into reality. We have finite capacities of perception and memory. But we are embedded in an infinite network of conscious agents whose complexity exceeds our finite capacities. So our interface must ignore all but a sliver of this complexity. For that sliver, it must deploy its capacities judiciously—more detail here, less there, next to nothing elsewhere. Hence our decline of insight as we shift our gaze from human to ant to quark. Our decline of insight should not be mistaken for an insight into decline—a progressive poverty inherent in objective reality. The decline is in our interface, in our perceptions. But we externalize it; we pin it on reality. Then we erect, from this erroneous reification, an ontology of physicalism.

Conscious realism pins the decline where it belongs—on our interface, not on an unconscious objective reality. Although each successive icon, in the sequence from human through ant to quark, offers a dimmer view of the conscious world that lies behind, this does not entail that consciousness itself is on a dimmer switch. The face I see in a mirror, being an icon, is not itself conscious. But behind that icon flourishes, I know firsthand, a living world of conscious experiences. Likewise, the stone I see in a riverbed, being an icon, is not conscious nor inhabited by consciousness. It is a pointer to a living world of conscious experiences no less vibrant than my own—just far more obscured by the limitations of my icon. Such limitation is to be expected in the perceptions of any finite creature facing a reality that, in comparison to itself, is infinitely complex.

I have touted the virtue of precision in a theory of consciousness. It's time to add some precision to the theory of conscious agents. Let's leave the mathematical definition of a conscious agent to the appendix. But behind the mathematical definition are simple intuitions.

Figure 42, from a few pages earlier, depicts two agents. Each agent has a set of possible experiences and a set of possible actions, and each agent perceives, decides, and acts. Each action is followed by an experience, perhaps desirable or perhaps not. Steal a carcass from lions: experience suffering. Pick a fig: experience a treat. Each action is a bet on future experiences. Sometimes you bet on a meal or a mate. Sometimes you bet your life.

To bet wisely, you must know the menu of options. At a horse race, for instance, your options might include picking Seabiscuit to show, place, or win; or hazarding a trifecta with Seabiscuit first, Secretariat second, and Big Red third.

A conscious agent needs a menu of actions, and a menu of the experiences that may follow. In mathematics, such a menu is called a *measurable space.*[16] It is the minimal structure you need to discuss probabilities, such as the probability that Seabiscuit will win. So the menus of actions and experiences of a conscious agent are measurable spaces. That's it. Nothing else. This is the minimal structure required to allow the theory of conscious agents to be testable by experiments.[17] If we could not describe probabilities of experiences and actions, we could not make empirical predictions from the theory. We could not do science.

A conscious agent is dynamic: it perceives, decides, and acts. When it perceives, its experience often changes; when it decides, its action often changes; when it acts, the experiences of other agents often change. Dynamics is conditional change. I see a blueberry muffin and butter croissant, and decide on the croissant; then I discover, behind the muffin, a chocolate eclair, and happily capitulate. My change in action, croissant to eclair, is a conditional change: it depends on my new experience, my tempting vision of a chocolate delight. Each new experience invites a new plan of action. In mathspeak, such a conditional change is a *Markovian kernel.*[18] The dynamics of a conscious agent—perceive, decide, and act—is, in each case, a Markovian kernel. Again, that's it.

In sum, a conscious agent has experiences and actions, which are menus (measurable spaces). It perceives, decides, and acts, which are conditional changes (Markovian kernels). And it counts how many experiences it has had. That's the entire definition of a conscious agent. It is, a mathematician would assure you, a simple bit of math.

"But," you might object, "this math can also describe mechanical agents that are unconscious. So it says nothing about consciousness."

This objection is a simple mistake. It's like saying that numbers can count apples and so they can't count oranges. Measurable spaces can describe unconscious events, such as flips of a coin. But they can also describe conscious events, such as experiences of taste and color. Probabilities and Markovian kernels can describe blind chance and unconscious decision, but also free will and conscious deliberation.

The definition of a conscious agent is just math. The math is not the territory. Just as a mathematical model of weather is not, and cannot create, blizzards and droughts, so also the mathematical model of conscious agents is not, and cannot create, consciousness. So, with this proviso, I offer a bold thesis, the *Conscious Agent Thesis*: every aspect of consciousness can be modeled by conscious agents.[19]

The definition of conscious agent is precise, and this thesis is bold—not because I know it is right, but because I want to discover where, precisely, it may be wrong and, if possible, to repair the defect. This is standard procedure in science: present a clear theory, paint a big target, and hope that gifted colleagues will try, by logic and experiment, to shoot it down. Where a shot hits the mark, try to improve the theory.

A theory must suffer the slings and arrows of opponents, but it also needs proponents. Here are some virtues of conscious agents. They are computationally universal: networks of conscious agents can perform any cognitive or perceptual task, including learning, memory, problem solving, and object recognition.[20] Several such networks have been constructed, and offer an alternative to traditional neural networks.[21] Conscious agents offer a promising new framework for the construction of theories in cognitive neuroscience. This framework does not assume that biological neurons and their networks are the building blocks of cognition. Instead it takes consciousness as fundamental and then has the task of showing how spacetime, matter, and neurobiology can emerge as components of the perceptual interface of certain conscious agents.

Conscious agents can combine to form new conscious agents, and these new agents can again combine to form yet higher agents, ad infinitum. When two or more agents interact, each retains its individual agency, but together they also instantiate a new agent. The more each of the agents in an interaction can predict its experiences from its actions, the more integrated is their joint dynamics and the more cohesive is the new agent that they instantiate. The decisions and actions of a higher-level agent can, in turn, influence the dynamics of the agents in its instantiation.

The decisions of a conscious agent have a contribution by that agent at its own level, plus contributions from the decisions of the agents in its instantiation. The decisions of an agent at its own level may correspond to Daniel Kahneman's "System 2" decisions, which are explicit and effortful, and the decisions further down in its instantiation may correspond to Kahneman's "System 1" decisions, which appear more emotional, attitudinal, and automatic.[22]

Combining agents into more complex agents can proceed ad infinitum, but unpacking agents into systems of simpler agents cannot. There is a bottom to the hierarchy of conscious agents. At the bottom reside the most elementary agents—"one-bit" agents—having just two experiences and two actions. The dynamics of a one-bit agent, and of interactions between two such agents, can be analyzed completely.[23] Here, at the foundation of agents, we can hope to connect with the foundations of spacetime, with physics at the Planck scale, and discern just how agents boot up a spacetime desktop.

The interface theory of perception contends that there is a screen—an interface—between us and objective reality. Can we hope to pierce that screen and see objective reality? Conscious realism says yes: we have met reality and it is like us. We are conscious agents, and so is objective reality. Beyond the interface lurks no Kantian noumenon, forever alien and impervious to our inquiry. Instead, we find agents like us: conscious agents. Their variety dwarfs the dazzling diversity of creatures that have paraded the earth and bequeathed to its sediments innumerable petrified mementos of their sojourn. We cannot imagine, concretely, even one new color. We cannot hope to imagine but a fraction of the varied experiences enjoyed by this multifarious host of agents. But despite our diversity, we share a unity: we are all agents, conscious agents.

"But," you might object, "didn't you earlier define 'objective reality' as that which exists even when no one observes? And don't conscious experiences exist only when some agent observes? Haven't you contradicted yourself when you propose conscious realism, and claim that objective reality consists of conscious agents?"

Indeed, for sake of argument, I adopted a notion of objective reality that is accepted by most physicalists. Then I used evolutionary assumptions that are also accepted by most physicalists to make the case against physicalism and its notion of objective reality. Now that I have presented that case, I am proposing a new ontology, and with it a new notion of objective reality in which conscious agents, with their experiences and structures, are central.

Conscious realism says that, despite our limits of imagination, a science of objective reality, of conscious agents and their interactions, is indeed possible. We can concretely imagine a space with at most three dimensions, but scientific theories routinely employ spaces with more dimensions, spaces that stump our imagination. In like manner, we can concretely imagine conscious experiences only within the tiny repertoire of *Homo sapiens*, but we can devise a scientific theory of all conscious agents, including those whose experiences stump our concrete imagination.

ITP and conscious realism reframe the classic problem of the relation between the brain and conscious experience. In chapter one, we discussed patients with split brains. When Joe Bogen severed a corpus callosum, his scalpel divided a unified brain into uncoupled hemispheres. This is a description of his surgery in the physicalist parlance of our interface. In reality, according to conscious realism, his scalpel split a conscious agent into two agents. The rich interactions of those two agents, which had instantiated a higher agent, became meager. We have seen that our interface can sometimes grant crude insight into the conscious realm behind—a smile can tell of joy, a deadpan tone of sorrow. Here, with its icon of a brain, our interface offers crude insight into agents and their combination—two lumps of meat joined by a corpus callosum tell of two agents interacting to form a new agent; two lumps with a severed callosum tell of an erstwhile unified agent now divorced into two distinct agents.

As we peer more closely at each hemisphere, our interface shows us networks of billions of neurons—again, perhaps granting crude insight into a realm of conscious agents that interact and instantiate higher agents. When we peer further into each neuron, and then into its chemistry, and finally into its physics, crude insight lapses into none.

A neuroscientist might object. "Cognitive neuroscience reveals that the vast majority of our mental processes are unconscious. We are unaware of the sophisticated processes by which we understand and produce speech, make decisions, learn, walk, understand, or transform images at the eye into visual worlds. Surely this vast swath of unconscious processing contradicts the claim of conscious realism that reality consists entirely of conscious agents. Conscious realism shipwrecks on the shoal of unconscious processes."

But this again mistakes a limit of our interface for an insight into reality. When I talk with a friend, I assume that she is conscious. I cannot directly experience her consciousness. It is inaccessible to me, and I can at best infer what it might be like to be her. But I would be mistaken to conclude that, because I am not conscious of her consciousness, she must be unconscious. Similarly, I would be mistaken to conclude that, because I am not conscious of some of my own mental processes, those processes must be unconscious. I can be unaware of many of my own mental processes, and yet those processes could themselves be conscious to other agents in my instantiation.

A conscious agent enjoys a repertoire of experiences. It networks with many other agents, which enjoy a stupefying variety of disparate repertoires. So it cannot experience the vast majority of these exotic experiences. This holds in particular for the hierarchy of agents that constitute its own instantiation. An agent simply lacks the resources to experience all the experiences of all the agents in its instantiation, even though those agents contribute to its very self. An agent can at best wield its repertoire of experiences to paint, with broad brush, a crude depiction of its instantiation. In our case, we paint a body, brain, neurons, chemicals, and particles on a canvas of spacetime. Then we step back, admire our handiwork, and conclude that there's nothing conscious to see here—a simple mistake that fosters physicalism and turns the problem of consciousness into a mystery.

A conscious agent is not just a repertoire of experiences. It decides and acts. But its actions are, by its very definition, distinct from its experiences: the diagram of an agent, for instance, has one box for "Experiences" and a separate box for "Actions." This entails that a conscious agent can be aware, and yet not self-aware—not aware of its own decisions and actions. To be aware of itself, an agent must devote some of its experiences, some of its perceptual interface, to represent some of its own decisions and actions. Its interface must have an icon, or icons, that represent the decisions and actions of the agent itself. If it sees itself at all, it sees itself through its own interface—as through a glass, darkly. And, of necessity, incompletely.

No conscious agent can describe itself completely. The very attempt adds more experiences to the agent, which multiplies the complexity of its decisions and actions in light of those new experiences, which requires yet more experiences to capture those more complex decisions and actions, and so on in a vicious loop of incompleteness. A conscious agent must therefore remain, at least in part, unconscious to itself. Recall that what conscious realism claims to be fundamental is not just conscious experiences, but conscious *agents*. An agent cannot experience itself in its entirety, no matter how large its repertoire of experiences. From this limitation may arise philosophical conundrums, personal angst, and job security for psychotherapists.

There is, however, good reason to fabricate a self. If you experience your acts and their consequences, then you can learn. If *this* act leads to *that* noxious experience, then you can learn not to do this act. The richer your experience of your internal decisions and actions, the more latitude you have for nuanced interactions with the outside world. To know other agents, you must also know yourself. All knowledge is, in this sense, embodied.

Conscious realism must pay another promissory note. It must, from first principles, describe precisely the dynamics of conscious agents, and show how this dynamics, when projected into the interface of *Homo sapiens*, appears as modern physics and Darwinian evolution. This is a strong empirical constraint on a theory of agent dynamics: its projection into our spacetime interface must account for all the data that supports modern physics and evolution. In addition, it must make new predictions that can be tested by experiments.

What principles, and dynamics of agents, might fill the bill? I'm not yet sure. But a tantalizing thread stretches from conscious agents through natural selection to physics. A basic law of physics says, informally, that everything falls apart. As the poet William Drummond (1585–1649) put it, "all beneath the moon decays, And what by mortals in this world is brought, In Time's great periods shall return to nought." More precisely, this law—the second law of thermodynamics—says that the total entropy of any isolated system never decreases. The rot of entropy is an implacable enemy of life, a purveyor of decay and death. Life, as evolutionary psychologists John Tooby, Leda Cosmides, and Clark Barrett explain, has but one defense: "natural selection is the only known natural process that pushes populations of organisms thermodynamically uphill into higher degrees of functional order, or even offsets the inevitable increase in disorder that would otherwise take place."[24]

Entropy is the information you lack—the number of yes-no questions you would need, as when playing the parlor game of Twenty Questions, to fill in what you don't know. But information, transacted in the currency of conscious experiences, is also the fungible commodity of conscious agents. Perhaps the dynamics of conscious agents is similar to the dynamics of cryptocurrencies, but with conscious experiences as the coin of the realm; enforcement of no double spending, when projected into the spacetime interface of *Homo sapiens*, might appear as a conservation law of physics. Or perhaps, as the physicist and inventor Federico Faggin has proposed, a central goal of conscious agents is mutual comprehension.[25] If so, then the dynamics of conscious agents may favor interactions that increase mutual information, and this dynamics, when projected from networks of agents into the interface of *Homo sapiens*, may appear there as evolution by natural selection. These are intriguing directions for research that may link insights from the theory of social networks—which describes why Google gets more hits than Hoffman—to the emergence of fitness functions in evolutionary biology.

Conscious realism advances an ontology radically different from the physicalism that dominates modern neuroscience, and science more generally. Radically different, but not radically new. Many key ideas of conscious realism and the interface theory of perception have appeared in prior sources, from ancient Greek philosophers such as Parmenides, Pythagoras, and Plato through more recent German philosophers such as Leibniz, Kant, and Hegel, and from eastern religions such as Buddhism and Hinduism to mystical strands of Islam, Judaism, and Christianity. The British philosopher and bishop George Berkeley clearly summarized some of the key ideas: "For as to what is said of the absolute existence of unthinking things without any relation to their being perceived, that seems perfectly unintelligible. Their ESSE is PERCIPI, nor is it possible they should have any existence out of the minds or thinking things which perceive them."[26]

If conscious agents and conscious realism contribute something new, it's to assemble old ideas from philosophy and religion into a theory of consciousness that is precise and testable. This allows the ideas to be refined under the watchful eye of the scientific method.

Science, like philosophy and religious practice, is a human endeavor. It is not infallible. Each of the many attempts to demarcate, from first principles, science from pseudoscience remains, at best, controversial.[27] What science offers is not gold-standard beliefs, but a potent method for winnowing beliefs that derives its power from the way it engages with human nature. We are a species that argues. Experiments show, and evolutionary theory explains, that we reason best when we argue for an idea that we already believe, or against the idea of another that we disbelieve.[28] We did not evolve our ability to reason in order to pursue the truth. We evolved it as a tool of social persuasion. As a result, our reasoning is plagued with foibles, such as a bias toward information that supports what we already believe. The scientific method exploits all of this. Each scientist argues for her idea, and against contradictory ideas of other scientists. In this argumentative context, our reason is at its sharpest: each idea garners the best support of reason and evidence its proponents can muster, and each endures the best impalement by reason and evidence its detractors can counter. Add to this sharpening of reason the demand that ideas be precise—mathematically precise, when possible—and the phoenix of science arises from foibles of human nature.

Science is not a theory of reality, but a method of inquiry. It orchestrates the better angels of our nature to promote reason, precision, productive dialog, and an appeal to evidence. It curbs our proclivity for the vague, deceptive, dogmatic, and imperious. Inquiry into any question that captures the human imagination—including meaning, purpose, values, beauty, and spirituality—deserves no less than the full benefit of this orchestration. Why deny ourselves our best chance to better understand?

Scholars of stature in science and religion have argued sometimes to the contrary. The US National Academy of Sciences, in its 1999 publication *Science and Creationism*, proposed that "Science tries to document the factual character of the natural world, and to develop theories that coordinate and explain these facts. Religion, on the other hand, operates in the equally important, but utterly different, realm of human purposes, meanings, and values—subjects that the factual domain of science might illuminate, but can never resolve." The evolutionary biologist Stephen Jay Gould likewise claimed that "science and religion occupy two separate realms of human experience. Demanding that they be combined detracts from the glory of each."[29]

Richard Dawkins disagreed, arguing that "it is completely unrealistic to claim, as Gould and many others do, that religion keeps itself away from science's turf, restricting itself to morals and values. A universe with a supernatural presence would be a fundamentally and qualitatively different kind of universe from one without. The difference is, inescapably, a scientific difference. Religions make existence claims, and this means scientific claims."[30]

I agree with Dawkins. If a system of thought, religious or otherwise, offers a claim that it wants taken seriously, then we should examine it with our best method of inquiry—the scientific method. That is taking it seriously.

Some topics—such as God, the good, reality, and consciousness—have been claimed to transcend the limited scope of human concepts and thus the methods of science. I have no quarrel with someone who claims this and then, being consistent, says no more about these topics. But if one does say more, then "What can be said at all can be said clearly" and probed with the scientific method. Can science describe who we are? I think so, in the sense that we can, by the scientific method, evolve and refine theories of who we are. But if science cannot describe who we are, then imprecise natural languages such as English certainly cannot describe who we are. We have no better means of crafting explanations than the scientific method. An explanation that descended from on high, but could not be tested and debated, would be no explanation at all.

"But," you might object, "the study of consciousness requires first-person experience. So it eludes science, which requires objective data obtained from a third-person point of view."

This claim is mistaken. Science is not an ontology. It is not committed to a spacetime and objects that existed before any first-person experiences, and that must be studied from a third-person stance. Science is a method. It can test and discard ontologies. If our perceptions evolved by natural selection, then, according to the FBT Theorem, we should discard the ontology of physicalism. We should recognize that spacetime and objects are the perceptual interface used by *Homo sapiens*. They are our first-person experiences. The scientific study of physical objects in spacetime, even when conducted by large teams of scientists using advanced technologies, is necessarily a study of first-person experiences.

The moon I see is an icon of my interface, and the moon you see is an icon of your interface. There is no objective moon or spacetime that exists even when unperceived and that must therefore be examined from a third-person point of view. There are only first-person observations. But they do not elude science. They are the only data science ever had. Science compares first-person observations to see if they agree. If they do, then we gain confidence in our observations and the theories they support. But each physical object we study by experiment is just an icon in an interface, not an element of objective reality beyond that interface. Intersubjective agreement about a physical object or a meter reading does not entail that the object or reading exist when no one observes.

Conscious realism makes a bold claim: consciousness, not spacetime and its objects, is fundamental reality and is properly described as a network of conscious agents.[31] To earn its keep, conscious realism must do serious work ahead. It must ground a theory of quantum gravity, explain the emergence of our spacetime interface and its objects, explain the appearance of Darwinian evolution within that interface, and explain the evolutionary emergence of human psychology.

Conscious realism offers a fresh take on a sci-fi motif: Can artificial intelligence (AI) create real consciousness? Physicalists assume that fundamental particles are not conscious, but some conjecture that an object—a system of insentient particles—can generate consciousness if its internal dynamics instantiates the right complexity. Sophisticated AI can ignite real consciousness.

Conscious realism contends, to the contrary, that no physical object is conscious. If I see a rock, then that rock is part of my conscious experience, but the rock itself is not conscious. When I see my friend Chris, I experience an icon that I create, but that icon itself is not conscious. My Chris-icon opens a small portal into the rich world of conscious agents; a smiling icon, for instance, suggests a happy agent. When I see a rock, I also interact with conscious agents, but my rock-icon offers no insight, no portal, into their experiences.

So conscious realism reframes the AI question: Can we engineer our interface to open new portals into the realm of conscious agents? A hodgepodge of transistors affords no insight into that realm. But can transistors be assembled and programmed into an AI that opens a new portal into that realm? For what it's worth, I think so. I think that AI can open new portals into consciousness, just as microscopes and telescopes open new vistas within our interface.

I also think that conscious realism can breach the wall between science and spirituality. This ideological barrier is a needless illusion, enforced by hoary misconceptions: that science requires a physicalist ontology that is anathema to spirituality, and that spirituality is impervious to the methods of science. I see ahead an uneasy truce and eventual rapprochement. Scientists won't readily trade physicalism for conscious realism. Religious devotees will hesitate to demote ancient texts from citadels of authority to fallible founts of inspiration, and to embrace the iconoclastic debates and meticulous experiments of the scientific method. But in the end, both will recognize that they lost nothing of value, and in return secured a cleaner shot at our biggest questions: Who are we? Where are we? And what are we in the world for?

I mentioned that conscious agents combine to create more and more complex agents. This process eventuates in infinite agents, with infinite potential for experiences, decisions, and actions. The idea of an infinite conscious agent sounds much like the religious notion of God, with the crucial difference that an infinite conscious agent admits precise mathematical description. We can prove theorems about such agents and their relationship to finite agents such as us. In the process we can foster what might be called a scientific theology, in which mathematically precise theories of God can be evolved, sharpened, and tested with scientific experiments. I suspect, for instance, that an infinite conscious agent is not omniscient, omnipotent, omnipresent, or alone in its infinity. Scientific theology is not Promethean poaching in the sacrosanct property of ancient religions; it is applying our best cognitive and experimental tools to our dearest questions. The abstract discoveries of scientific theology would need to be translated into practical applications for laypersons. Religion can become an evolving science—informed by cognitive neuroscience and evolutionary psychology—whose salutary application to daily life also evolves.

The theory of God that emerges from a scientific theology need not posit a magician that flouts the laws of physics. These laws do not describe an unconscious reality; they describe the dynamics of conscious agents, finite and infinite, projected into the language and data structures of the spacetime interface of *Homo sapiens*. The laws of physics do not describe a machine, in which a marginalized ghost of consciousness must perform paranormal tricks to prove its existence. Consciousness need not flout scientific laws that are themselves projected descriptions of the dynamics of consciousness.

Suppose you drive with friends to a virtual-reality arcade to play volleyball. You slip on headsets and body suits, and find your avatars clad in swimsuits, immersed in sunshine, standing on a sandy beach with a volleyball net, surrounded by swaying palms and crying gulls. You serve the ball and start playing with abandon. After a while, one of your friends says he's thirsty and will be right back. He slips out of his headset and body suit. His avatar collapses onto the sand, inert and unresponsive. But he's fine. He just stepped out of the virtual-reality interface.

When we die, do we simply slip out of the spacetime interface of *Homo sapiens*? I don't know. But we have the theory of conscious realism, and the mathematics of conscious agents. Let's do some science.

Conscious realism claims that consciousness is the fundamental nature of objective reality. I have been warned that this is an anachronism that misses the key message of the Copernican revolution: it's not about us. We used to think that everything is about us and that therefore the earth must be the center of the universe. When Copernicus and Galileo discovered that it isn't, this forced us to adjust our astronomy, but more importantly it forced us to transform our conception of ourselves. We are not center stage. We cling to a tiny rock in a nondescript corner of a vast universe. We aren't even bit players. And this, I have been told, is what conscious realism gets wrong. By placing consciousness at the center of reality, conscious realism tries to return to a pre-Copernican era in which we could naïvely believe that we, and our consciousness, are the raison d'être of the universe.

This critique misreads conscious realism. It claims no central role for human consciousness. It posits countless kinds of conscious agents with a boundless variety of conscious experiences, most of which we cannot concretely imagine. There is nothing special or central about human beings as conscious agents. To say that consciousness is fundamental is not to say that human consciousness is fundamental or distinctive.

This critique also misreads the Copernican revolution. Yes, our perceptions misled us about our place in the universe. But its deeper message is this: our perceptions can mislead us about the very nature of the universe itself. We are prone to falsely believe that certain limitations and idiosyncrasies of our perceptions are genuine insights into objective reality. Galileo got the message and fingered some culprits. "I think that tastes, odors, colors, and so on . . . reside in consciousness. Hence if the living creature were removed, all these qualities would be wiped away and annihilated." Galileo denied that our perceptions of tastes, odors, and colors are genuine insights into objective tastes, odors, and colors. There are, he claimed, no tastes, odors, or colors in objective reality. These are just features of our perceptions.

Galileo got the message, took a giant leap in the right direction, and then stopped. He still held that our perceptions of objects in space, with their shapes, positions, and momenta, are genuine insights into the true nature of objective reality. Most of us would agree.

But the theory of evolution by natural selection disagrees. It declares that the Copernican revolution extends farther than Galileo imagined. Objects, shapes, space, and time reside in consciousness. If the living creature were removed, all these qualities would be annihilated. Physics does not demur. Indeed, physicists concede that spacetime is doomed. It is not the primordial stage on which the drama of life plays out.

What is spacetime? This book has offered you the red pill. Spacetime is your virtual reality, a headset of your own making. The objects you see are your invention. You create them with a glance and destroy them with a blink.

You have worn this headset all your life. What happens if you take it off?

APPENDIX

Precisely

The Right to Be Wrong

This brief appendix presents the mathematical definition of a conscious agent. Conscious agents can form networks to perform any cognitive task. For those wanting more details, several papers develop the properties of conscious agents and their applications.[1]

DEFINITION. A *conscious agent*, C, is a seven tuple $C = (X, G, W, P, D, A, T)$, where X, G, and W are measurable spaces, $P: W \times X \rightarrow X$, $D: X \times G \rightarrow G$, and $A: G \times W \rightarrow W$ are Markovian kernels,[2] and T is a totally ordered set.

The space X of a conscious agent represents its possible conscious experiences, G its possible actions, and W the world. The perception kernel P describes how the state of the world influences its state of perception; the decision kernel D describes how the state of its perception influences its choice of action; and the action kernel A describes how its action influences the state of the world. The counter T increments with each new decision of the conscious agent. The requirement that X, G, and W are measurable spaces is made to allow the use of probabilities and probabilistic predictions, which are essential to science. This requirement can be relaxed, without losing probabilistic prediction: σ-algebras, which are closed under countable union, can be relaxed to finite additive classes, which are closed under finite disjoint union.

Just as any effective computation can, according to the Church-Turing thesis, be couched in the formalism of a Turing machine, so also any aspect of consciousness and agency can, according to the conscious-agent thesis, be couched in the formalism of a conscious agent.[3] This is an empirical proposal that one can try to refute by counterexample. Conscious realism is the hypothesis that the world, *W*, is a network of interacting conscious agents.

Conscious agents can combine in several ways to form new, perhaps more complex, conscious agents.[4] For instance, because Markovian kernels can be composed to create a new, single Markovian kernel, the decision kernel of one conscious agent can be replaced by another entire conscious agent; and similarly for the perception and action kernels. This is possible because perception, decision, and action are each modeled as a Markovian kernel. Thus, although the basic definition of conscious agent may appear at first to put a strong divide between perceptions, decisions, and actions, in fact it allows for their intermixing.

Two agents, $C_1 = (X_1, G_1, W, P_1, D_1, A_1, T_1)$ and $C_2 = (X_2, G_2, W, P_2, D_2, A_2, T_2)$ that interact as depicted in Figure 42 combine to form a single agent. According to conscious realism, this entails that the interaction of any agent with the rest of the world can be modeled as a two-agent interaction. We can compress any two-agent interaction into $\mathcal{G}(2,4)$, the conformal geometric algebra for a spacetime with signature (1, 3). $\mathcal{G}(2,4)$ has a standard orthogonal basis $\{\gamma_0, \gamma_1, \gamma_2, \gamma_3, e, \bar{e}\}$, with $\gamma_0^2 = e^2 = 1$ and $\gamma_1^2 = \gamma_2^2 = \gamma_3^2 = \bar{e}^2 = -1$; it has graded subspaces of dimensions 1, 6, 15, 20, 15, 6, and 1. Its rotor group is isomorphic to the Lie group SU(2,2).[5]

For two finite agents whose measurable spaces each have cardinality *N*, we order the elements of each measurable space, and associate to each element its index in this arbitrary but fixed order. We let $t_1 \in \{0, ..., N-1\}$ denote the index of an element of T_1; we let t_2 denote the index of an element in T_2; and similarly, mutatis mutandis, for x_1, g_1, x_2 and g_2. Then we can map this pair of agents and its dynamics into a discrete spacetime using the mapping $\kappa : X_1 \times G_1 \times T_1 \times X_2 \times G_2 \times T_2 \to \mathcal{G}(2,4)$ given by $(x_1, g_1, t_1, x_2, g_2, t_2) \mapsto t_1\gamma_0 + t_2 e + x_1\gamma_1 + g_1\gamma_2 + x_2\gamma_3 + g_2\bar{e}$. Here the geometric algebra is over the ring $\mathbb{Z}^N$. The map κ takes T_1 into γ_0, X_1 into γ_1, G_1 into γ_2, T_2 into e, X_2 into γ_3, G_2 into $\bar{e}$ and induces a compression of the Markovian dynamics of conscious agents into a spacetime dynamics. Thus is a fundamental bridge between the objective reality of interacting conscious agents and the representation of that reality in a spacetime interface of some conscious agent, say agent C_1. If this interface occupies a subset of X_1, and if X_1 has cardinality N then its representation of $\mathcal{G}(2,4)$ must be over a ring $\mathbb{Z}^M$, with $M < N$; in fact, M must be substantially smaller than N. This case is necessarily self-referential, because γ_0, γ_1, γ_2 and represent respectively T_1, X_1, and G_1.

A simple network is a pair of "one-bit" conscious agents, for which $N = 2$. Its compression into a discrete spacetime may correspond to the Planck scale. Two one-bit agents can combine to comprise a two-bit agent, for which $N = 4$. A pair of two-bit agents have a compression into spacetime that is richer than the one-bit case. Two two-bit agents can combine to comprise a four-bit agent, and so on ad infinitum. In the limit we approach a continuous spacetime representation. In this process, we compress the infinite complexity of the network of conscious agents into a spacetime data format. The network dynamics of conscious agents is compressed into dynamics within spacetime. For instance, perhaps a dynamical evolution of conscious agents toward small-world networks may appear in spacetime as the dynamics of gravity.[6]

ACKNOWLEDGMENTS

Research is inspired by island hopping in the archipelago of human knowledge. With luck, you discover new outcrops, and tantalizing hints of ecosystems offshore and continents beyond.

Tips from fellow explorers have been a great help. For sharing their insights, I thank Chris Anderson, Patrick Bender, Jordan Biren, Erie Boorman, Lindsay Bowman, Kees Brouwer, Andrew Burton, August Bradley Cenname, David Chalmers, Deepak Chopra, Annie Day, Dan Dennett, Jochen Ditterich, Zoe Drayson, Mike D'Zmura, Federico Faggin, Chris Fields, Scott Fisher, Pete Foley, Joy Geng, Greg Hickok, Perry Hoberman, David and Loretta Hoffman, Eve Isham, Petr Janata, Greg Kendall, Virginia Kuhn, Steve Luck, Brian Marion, Justin Mark, Andrew McNeely, Lee Miller, Jennifer Moon, Louis Narens, Darren Peshek, Steven Pinker, Zygmunt Pizlo, Chetan Prakash, Robert Prentner, V. S. Ramachandran, Don Saari, Manish Singh, and Jörg Wallaschek.

In 2015, key ideas of this book appeared in "The interface theory of perception," a paper that I wrote with Manish Singh and Chetan Prakash, and published in a special issue of the journal *Psychonomic Bulletin & Review.* Several thoughtful commentaries appeared with the paper. For these I thank Bart Anderson, Jonathan Cohen, Shimon Edelman, Jacob Feldman, Chris Fields, E. J. Green, Greg Hickok, John Hummel, Scott Jordan, Jan Koenderink, Gary Lupyan, Rainer Mausfeld, Brian McLaughlin, Zygmunt

Pizlo, and Matthew Schlesinger. My thanks to Greg Hickok, who orchestrated the special issue and edited our paper.

Some friends, students, and colleagues went the extra mile, commenting on earlier drafts. For this, I thank Rugero Altair, Chris Anderson, Emma Brant, Andrew Burton, Deepak Chopra, Coleman Dobson, Maziar Esfahanian, Federico Faggin, Chris Fields, Pete Foley, Max Jones, Greg Kestin, Jack Loomis, Erin McKeon, Chetan Prakash, Robert Prentner, Rob Reid, Jenessa Reyes, Manish Singh, Tony Sobrado, Matthew Tillis, Janelle Vo, Mike Webster, and Emily Wong.

Special thanks to my agents, John Brockman and Katinka Matson, who encouraged me to take on this project, and to Max Brockman for negotiating with publishers. Special thanks also to Quynh Do, my editor at Norton, who streamlined my prose and made key concepts more accessible.

The Case Against Reality includes several depictions of the Subjective Necker Cube image (one of which appears on the cover of the original hardcover edition) that were attributed to me, to which Dr. Heywood Petry objected. The Subjective Necker Cube image was originally created by Dr. Heywood Petry. Dr. Petry's Subjective Necker Cube image is widely recognized by vision scientists as an inspired illustration and demonstration of the human visual capacity to create subjective contours and objects in three dimensions. Any third-party requests for permission to use the Subjective Necker Cube image should be directed to Dr. Petry.

My work on this book was facilitated by a sabbatical leave granted by the University of California–Irvine, and by generous gifts from the Federico and Elvia Faggin Foundation. I am most grateful.

Loving thanks to my wife, Geralyn Souza, for her encouragement, patience, and love throughout.

NOTES

Preface

1. Taylor, C. C. W. 1999. "The atomists," in A. A. Long, ed., *The Cambridge Companion to Early Greek Philosophy* (New York: Cambridge University Press), 181–204, doi: 10.1017/CCOL0521441226.009.
2. Plato, *The Republic*, Book VII.
3. *Spacetime* is thus a technical term from physics. I will use it when emphasizing technical issues from physics and information theory. I will use "space" and "time" separately when emphasizing them as separate aspects of our perceptual experiences.
4. Chamovitz, D. 2012. *What a Plant Knows* (New York: Scientific American / Farrar, Straus and Giroux).
5. Wiltbank, L. B., and Kehoe, D. M. 2016. "Two cyanobacterial photoreceptors regulate photosynthetic light harvesting by sensing teal, green, yellow and red light," mBio 7 (1): e02130-15, doi: 10.1128/mBio.02130-15.
6. This refers to *The Matrix*, a movie in which the protagonist's choice between a red pill and blue pill alters his fate.

Chapter One: Mystery

1. Bogen, J. 2006. "Joseph E. Bogen," in L. Squire, ed., *The History of Neuroscience in Autobiography, Volume 5* (Amsterdam: Elsevier), 47–124.
2. Leibniz, G. W. 1714/2005. *The Monadology* (New York: Dover).
3. Huxley, T. 1869. *The Elements of Physiology and Hygiene: A Text-book for Educational Institutions* (New York: Appleton), 178.
4. James, W. 1890. *The Principles of Psychology* (New York: Henry Holt), 1:146, 147.
5. Freud, S. 1949. *An Outline of Psycho-Analysis*, trans. J. Strachey (London: Hogarth Press), 1.
6. Crick, F. 1994. *The Astonishing Hypothesis* (New York: Scribner's), 3.

7. Sperry, R.W. 1974. "Lateral specialization of cerebral function in the surgically separated hemispheres," in R. McGuigan and R. Schoonover, eds., *The Psychophysiology of Thinking* (New York: Academic Press), 213.
8. Ledoux, J. E., Wilson, D. H., and Gazzaniga, M. S. 1977. "A divided mind: Observations on the conscious properties of the separated hemispheres," *Annals of Neurology* 2: 417–21.
9. https://www.youtube.com/watch?v=PFJPtVRlI64.
10. Desimone, R., Schein, S. J., Moran, J., and Ungerleider, L. G. 1985. "Contour, color and shape analysis beyond the striate cortex," *Vision Research* 25: 441–52; Desimone, R., and Schein, S. J. 1987. "Visual properties of neurons in area V4 of the macaque: Sensitivity to stimulus form," *Journal of Neurophysiology* 57: 835–68; Heywood, C. A., Gadotti, A., and Cowey, A. 1992. "Cortical area V4 and its role in the perception of color," *Journal of Neuroscience* 12: 4056–65; Heywood, C. A., Cowey, A., and Newcombe, F. 1994. "On the role of parvocellular (P) and magnocellular (M) pathways in cerebral achromatopsia," *Brain* 117: 245–54; Lueck, C. J., Zeki, S., Friston, K. J., Deiber, M.-P., Cope, P., Cunningham, V. J., Lammertsma, A. A., Kennard, C., and Frackowiak, R. S. J. 1989. "The colour centre in the cerebral cortex of man," *Nature* 340: 386–89; Motter, B. C. 1994. "Neural correlates of attentive selection for color or luminance in extrastriate area V4," *Journal of Neuroscience* 14: 2178–89; Schein, S. J., Marrocco, R. T., and de Monasterio, F. M. 1982. "Is there a high concentration of color-selective cells in area V4 of monkey visual cortex?" *Journal of Neurophysiology* 47: 193–213; Shapley, R., and Hawken, M. J. 2011. "Color in the cortex: Single- and double-opponent cells," *Vision Research* 51: 701–17; Yoshioka, T., and Dow, B. M. 1996. "Color, orientation and cytochrome oxidase reactivity in areas V1, V2, and V4 of macaque monkey visual cortex," *Behavioural Brain Research* 76: 71–88; Yoshioka, T., Dow, B. M., and Vautin, R. G. 1996. "Neuronal mechanisms of color categorization in areas V1, V2, and V4 of macaque monkey visual cortex," *Behavioural Brain Research* 76: 51–70; Zeki, S. 1973. "Colour coding in rhesus monkey prestriate cortex," *Brain Research* 53: 422–27; Zeki, S. 1980. "The representation of colours in the cerebral cortex," *Nature* 284: 412–18; Zeki, S. 1983. "Colour coding in the cerebral cortex: The reaction of cells in monkey visual cortex to wavelengths and colours," *Neuroscience* 9: 741–65; Zeki, S. 1985. "Colour pathways and hierarchies in the cerebral cortex," in D. Ottoson and S. Zeki, eds., *Central and Peripheral Mechanisms of Colour Vision* (London: Macmillan).
11. Sacks, O. 1995. *An Anthropologist on Mars* (New York: Vintage Books), 34.
12. Ibid., 28; Zeki, S. 1993. *A Vision of the Brain* (Boston: Blackwell Scientific Publications), 279.
13. Penfield, W., and Boldrey, E. 1937. "Somatic motor and sensory representation in the cerebral cortex of man as studied by electrical stimulation," *Brain* 60(4): 389–443.
14. Ramachandran, V. S. 1998. *Phantoms in the Brain* (New York: William Morrow).
15. Chalmers, D. 1998. "What is a neural correlate of consciousness?" in T. Metzinger, ed., *Neural correlates of consciousness: Empirical and conceptual questions* (Cambridge, MA: MIT Press), 17–40; Koch, C. 2004. *The Quest for Consciousness: A Neurobiological Approach* (Englewood, CO: Roberts & Company Publishers).

16. For more on the puzzles of causation, see Beebee, H., Hitchcock, C., and Menzies, P., eds. 2009. *The Oxford Handbook of Causation* (Oxford, UK: Oxford University Press).
17. Tagliazucchi, E., Chialvo, D. R., Siniatchkin, M., Amico, E., Brichant, J-F., Bonhomme, V., Noirhomme, Q., Laufs, H., and Laureys, S. 2016. "Large-scale signatures of unconsciousness are consistent with a departure from critical dynamics," *Journal of the Royal Society, Interface* 13: 20151027.
18. Chalmers, D. 1998. "What is a neural correlate of consciousness?" in T. Metzinger, ed., *Neural correlates of consciousness: Empirical and conceptual questions* (Cambridge, MA: MIT Press), 17–40; Koch, C. 2004. *The Quest for Consciousness: A Neurobiological Approach* (Englewood, CO: Roberts & Company Publishers).
19. Aru, J., Bachmann, T., Singer, W., and Melloni, L. 2012. "Distilling the neural correlates of consciousness," *Neuroscience and Behavioral Reviews* 36: 737–46.
20. Kindt, M., Soeter, M., and Vervliet, B. 2009. "Beyond extinction: Erasing human fear responses and preventing the return of fear," *Nature Neuroscience* 12(3): 256–58; Soeter, M., and Kindt, M. 2015. "An abrupt transformation of phobic behavior after a post-retrieval amnesic agent," *Biological Psychiatry* 78: 880–86.
21. Denny, C. A., et al. 2014. "Hippocampal memory traces are differentially modulated by experience, time, and adult neurogenesis," *Neuron* 83: 189–201; Cazzulino, A. S., Martinez, R., Tomm, N. K., and Denny, C. A. 2016. "Improved specificity of hippocampal memory trace labeling," *Hippocampus*, doi: 10.1002/hipo.22556.
22. Blackmore, S. 2010. *Consciousness: An Introduction* (New York: Routledge); Chalmers, D. 1996. *The Conscious Mind* (Oxford, UK: Oxford University Press); Revonsuo, A. 2010. *Consciousness: The Science of Subjectivity* (New York: Psychology Press).
23. One might object that the integrated information theory of Tononi does propose such laws (Oizumi, M., Albantakis, L., and Tononi, G. 2014. "From the phenomenology to the mechanisms of consciousness: Integrated information theory 3.0," *PLOS Computational Biology* 10: e1003588). But it does not. It gives no laws that identify specific conscious experiences, such as the taste of chocolate, with specific types of brain activity. And it gives no laws about how a specific kind of experience must change as the specific brain activity changes. The same is true of reductive functionalist theories of mind, which identify mental states (including conscious experiences) with functional processes of computational systems, whether biological or not. No reductive functionalist has proposed a single specific identity between a specific conscious experience (or a class of conscious experiences) and specific functional processes. Reductive functionalism has the further problem that, according to the Scrambling Theorem, it is provably false (Hoffman, D. D. 2006a. "The scrambling theorem: A simple proof of the logical possibility of spectrum inversion," *Consciousness and Cognition* 15: 31–45; Hoffman, D. D. 2006b. "The Scrambling Theorem unscrambled: A response to commentaries," *Consciousness and Cognition* 15: 51–53). The Scrambling Theorem also entails that conscious experiences are not identical to using information to perceive affordances and guide behavior in real time. Chemero (in Chemero, A. 2009. *Radical Embodied Cognitive Science* [Cambridge, MA: MIT Press]), for instance, claims that "In radical embodied cognitive science, using information

to perceive affordances and guide behavior in real time just is having conscious experiences. When we have explained how animals use information to directly perceive and act in their niches, we will also have explained their conscious experience." The Scrambling Theorem proves that this claimed identity is false. Moreover, no proponent of embodied cognition has ever proposed a single specific identity between a specific conscious experience (or a class of conscious experiences) and a specific use of information to perceive affordances and guide behavior in real time. Nor are there any proposals for principles that would explain such identities—why should a specific use of information to perceive affordances and guide behavior in real time be the conscious experience, say, of the taste of vanilla? Why couldn't that specific use of information to perceive affordances and guide behavior in real time be, say, the taste of chocolate or the feeling of a smooth cold column of ice? What scientific principles rule out the other conscious experiences? None have ever been offered. According to the Scrambling Theorem, there are no such principles.

24. Chomsky, N. 2016. *What Kind of Creatures Are We*? (New York: Columbia University Press).
25. Anscombe, G. E. M. 1959. *An Introduction to Wittgenstein's Tractatus* (New York: Harper & Row), 151.
26. Lovejoy, A. O. 1964. *The Great Chain of Being* (Cambridge, MA: Harvard University Press).
27. Galilei, G. 1623. *The Assayer*, trans. in Drake, S. 1957. *Discoveries and Opinions of Galileo* (New York: Doubleday), 274.

Chapter Two: Beauty

1. Gwynne, D. T., and Rentz, D. C. F. 1983. "Beetles on the Bottle: Male Buprestids Make Stubbies for Females," *Journal of Australian Entomological Society* 22: 79–80; Gwynne, D. T. 2003. "Mating mistakes," in V. H. Resh and R. T. Carde, eds., *Encyclopedia of Insects* (San Diego: Academic Press). Approximately one-quarter of all animal species on Earth are beetles (Bouchard, P., ed. 2014. *The Book of Beetles* [Chicago: University of Chicago Press]).
2. Wilde, O. 1894. *A Woman of No Importance* (London: Methuen, Third Act).
3. Langlois, J. H., Roggman, L. A., and Reiser-Danner, L. A. 1990. "Infants' differential social responses to attractive and unattractive faces," *Developmental Psychology* 26: 153–59.
4. Doyle, A. C. 1891/2011. *The Boscombe Valley Mystery* (Kent, England: Solis Press).
5. An image of Sharbat Gula can be seen at https://en.wikipedia.org/wiki/File:Sharbat_Gula.jpg.
6. Peshek, D., Sammak-Nejad, N., Hoffman, D. D., and Foley, P. 2011. "Preliminary evidence that the limbal ring influences facial attractiveness," *Evolutionary Psychology* 9: 137–46.
7. Ibid.
8. Peshek, D. 2013. "Evaluations of facial attractiveness and expression," PhD diss., University of California–Irvine.

9. Cingel, N. A. van der. 2000. *An Atlas of Orchid Pollination: America, Africa, Asia and Australia* (Rotterdam: Balkema), 207–8.
10. Gronquist, M., Schroeder, F. C., Ghiradella, H., Hill, D., McCoy, E. M., Meinwald, J., and Eisner, T. 2006. "Shunning the night to elude the hunter: Diurnal fireflies and the 'femmes fatales,'" *Chemoecology* 16: 39–43; Lloyd, J. E. 1984. "Occurrence of aggressive mimicry in fireflies," *Florida Entomologist* 67: 368–76.
11. Sammaknejad, N. 2012. "Facial attractiveness: The role of iris size, pupil size, and scleral color," PhD diss., University of California–Irvine.
12. Carcio, H. A. 1998. *Management of the Infertile Woman* (Philadelphia: Lippincott Williams & Wilkins); Rosenthal, M. S. 2002. *The Fertility Sourcebook*. 3rd edition (Chicago: Contemporary Books).
13. Buss, D. M. 2016. *Evolutionary Psychology: The New Science of the Mind*, 5th edition (New York: Routledge), Figure 5.1.
14. Kenrick, D. T., Keefe, R. C., Gabrielidis, C., and Cornelius, J. S. 1996. "Adolescents' age preferences for dating partners: Support for an evolutionary model of life-history strategies," *Child Development* 67: 1499–1511.
15. The ratio of iris width to eye width was .42 in one face and .48 in the other.
16. Sammaknejad, N. 2012. "Facial attractiveness: The role of iris size, pupil size, and scleral color," PhD diss., University of California–Irvine.
17. This was first proposed in Trivers, R. L. 1972. "Parental investment and sexual selection," in B. Campbell, ed. *Sexual Selection and the Descent of Man: 1871–1971*, 1st edition (Chicago: Aldine), 136–79. See also Woodward, K., and Richards, M. H. 2005. "The parental investment model and minimum mate choice criteria in humans," *Behavioral Ecology* 16(1): 57–61.
18. Trivers, R. L. 1985. *Social Evolution* (Menlo Park, CA: Benjamin/Cummings); but see Masonjones, H. D., and Lewis, S. M. 1996. "Courtship behavior in the dwarf seahorse *Hippocampus zosterae*," *Copeai* 3: 634–40.
19. Jones, I. L., and Hunter, F. M. 1993. "Mutual sexual selection in a monogamous seabird," *Nature* 362: 238–39; Jones, I. L., and Hunter, F. M. 1999. "Experimental evidence for a mutual inter- and intrasexual selection favouring a crested auklet ornament," *Animal Behavior* 57(3): 521–28; Zubakin, V. A., Volodin, I. A., Klenova, A. V., Zubakina, E. V., Volodina, E. V., and Lapshina, E. N. 2010. "Behavior of crested auklets (Aethia cristatella, Charadriiformes, Alcidae) in the breeding season: Visual and acoustic displays," *Biology Bulletin* 37(8): 823–35.
20. Smuts, B. B. 1995. "The evolutionary origins of patriarchy," *Human Nature* 6: 1–32.
21. Buss, D. M. 1994. "The strategies of human mating," *American Scientist* 82: 238–49; Gil-Burmann, C., Pelaez, F., and Sanchez, S. 2002. "Mate choice differences according to sex and age: An analysis of personal advertisements in Spanish newspapers," *Human Nature* 13: 493–508; Khallad, Y. 2005. "Mate selection in Jordan: Effects of sex, socio-economic status, and culture," *Journal of Social and Personal Relationships*, 22: 155–68; Todosijevic, B., Ljubinkovic, S., and Arancic, A. 2003. "Mate selection criteria: A trait desirability assessment study of sex differences in Serbia," *Evolutionary Psychology* 1: 116–26; Moore, F. R., Cassidy, C., Smith, M. J. L., and Perrett, D. I. 2006. "The effects of female control of resources on sex-differentiated mate preferences," *Evolution and Human Behavior* 27: 193–205; Lippa, R. A. 2009.

"Sex differences in sex drive, sociosexuality, and height across 53 nations: Testing evolutionary and social structural theories," *Archives of Sexual Behavior* 38: 631–51; Schmitt, D. P. 2012. "When the difference is in the details: A critique of Zentner and Mtura Stepping out of the caveman's shadow: Nations' gender gap predicts degree of sex differentiation in mate preferences," *Evolutionary Psychology* 10: 720–26; Schmitt, D. P., Youn, G., Bond, B., Brooks, S., Frye, H., Johnson, S., Klesman, J., Peplinski, C., Sampias, J., Sherrill, M., and Stoka, C. 2009. "When will I feel love? The effects of culture, personality, and gender on the psychological tendency to love," *Journal of Research in Personality* 43: 830–46.

22. Buss, D. M., and Schmitt, D. P. 1993. "Sexual strategies theory: An evolutionary perspective on human mating," *Psychological Review* 100: 204–32; Brewer, G., and Riley, C. 2009. "Height, relationship satisfaction, jealousy, and mate retention," *Evolutionary Psychology* 7: 477–89; Courtiol, A., Ramond, M., Godelle, B., and Ferdy, J. 2010. "Mate choice and human stature: Homogamy as a unified framework for understanding mate preferences," *Evolution* 64(8): 2189–2203; Dunn, M. J., Brinton, S., and Clark, L. 2010. "Universal sex differences in online advertisers' age preferences: Comparing data from 14 cultures and 2 religious groups," *Evolution and Human Behavior* 31: 383–93; Ellis, B. J. 1992. "The evolution of sexual attraction: Evaluative mechanisms in women," in J. Barkow, L. Cosmides, and J. Tooby, eds., *The Adapted Mind* (New York: Oxford), 267–288; Cameron, C., Oskamp, S., and Sparks, W. 1978. "Courtship American style: Newspaper advertisements," *Family Coordinator* 26: 27–30.
23. Rhodes, G., Morley, G., and Simmons, L. W. 2012. "Women can judge sexual unfaithfulness from unfamiliar men's faces," *Biology Letters* 9: 20120908.
24. Leivers, S., Simmons, L. W., and Rhodes, G. 2015. "Men's sexual faithfulness judgments may contain a kernel of truth," *PLoS ONE* 10(8): e0134007, doi: 10.1371/journal.pone.0134007.
25. Thornhill, R., Gangestad, S. W. 1993. "Human facial beauty: Averageness, symmetry and parasite resistance," *Human Nature* 4: 237–69; Thornhill, R., and Gangestad, S. W. 1999. "Facial attractiveness," *Trends in Cognitive Science* 3: 452–60; Thornhill, R., and Gangestad, S. W. 2008. *The Evolutionary Biology of Human Female Sexuality* (New York: Oxford University Press); Penton-Voak, I. S., Perrett, D. I., Castles, D. L., Kobayashi, T., Burt, D. M., Murray, L. K., and Minamisawa, R. 1999. "Female preference for male faces changes cyclically," *Nature* 399: 741–42.
26. Muller, M. N., Marlowe, F. W., Bugumba, R., and Ellison, P. T. 2009. "Testosterone and paternal care in East African foragers and pastoralists," *Proceedings of the Royal Society, B* 276: 347–54; Storey, A. E., Walsh, C. J., Quinton, R. L., and Wynne-Edwards, K. E. 2000. "Hormonal correlates of paternal responsiveness in new and expectant fathers," *Evolution and Human Behavior* 21: 79–95.
27. DeBruine, L., Jones, B. C., Frederick, D. A., Haselton, M. G., Penton-Voak, I. S., and Perrett, D. I. 2010. "Evidence for menstrual cycle shifts in women's preferences for masculinity: A response to Harris (in press), 'Menstrual cycle and facial preferences reconsidered,'" *Evolutionary Psychology* 8: 768–75; Johnston, V. S., Hagel, R., Franklin, M., Fink, B., and Grammer, K. 2001. "Male facial attractiveness: Evidence for a hormone-mediated adaptive design," *Evolution and Human Behavior* 22: 251–

67; Jones, B. C., Little, A. C., Boothroyd, L. G., DeBruine, L. M., Feinberg, D. R., Law Smith, M. J., Moore, F. R., and Perrett, D. I. 2005, "Commitment to relationships and preferences for femininity and apparent health in faces are strongest on days of the menstrual cycle when progesterone level is high," *Hormones and Behavior* 48: 283–90; Little, A. C., Jones, B. C., and DeBruine, L. M. 2008. "Preferences for variation in masculinity in real male faces change across the menstrual cycle," *Personality and Individual Differences* 45: 478–82; Vaughn, J. E., Bradley, K. I., Byrd-Craven, J., and Kennison, S. M. 2010. "The effect of mortality salience on women's judgments of male faces," *Evolutionary Psychology* 8: 477–91.

28. Johnston, L., Arden, K., Macrae, C. N., and Grace, R. C. 2003. "The need for speed: The menstrual cycle and personal construal," *Social Cognition* 21: 89–100; Macrae, C. N., Alnwick, K. A., Milne, A. B., and Schloerscheidt, A. M. 2002. "Person perception across the menstrual cycle: Hormonal influences on social-cognitive functioning," *Psychological Science* 13: 532–36; Roney, J. R., and Simmons, Z. L. 2008. "Women's estradiol predicts preference for facial cues of men's testosterone," *Hormones and Behavior* 53: 14–19; Rupp, H. A., James, T. W., Ketterson, E. D., Sengelaub, D. R., Janssen, E., and Heiman, J. R. 2009. "Neural activation in women in response to masculinized male faces: Mediation by hormones and psychosexual factors," *Evolution and Human Behavior* 30: 1–10; Welling, L. L., Jones, B. C., DeBruine, L. M., Conway, C. A., Law Smith, M. J., Little, A. C., Feinberg, D. R., Sharp, M. A., and Al-Dujaili, E. A. S. 2007. "Raised salivary testosterone in women is associated with increased attraction to masculine faces," *Hormones and Behavior* 52: 156–61.

29. Feinberg, D. R., Jones, B. C., Law Smith, M. J., Moore, F. R., DeBruine, L. M., Cornwell, R. E., Hillier, S. G., and Perrett, D. I. 2006. "Menstrual cycle, trait estrogen level, and masculinity preferences in the human voice," *Hormones and Behavior* 49: 215–22; Gangestad, S. W., Simpson, J. A., Cousins, A. J., Garver-Apgar, C. E., and Christensen, P. N. 2004. "Women's preferences for male behavioral displays change across the menstrual cycle," *Psychological Science* 15: 203–7; Gangestad, S. W., Garver-Apgar, C. E., Simpson, J. A., and Couins, A. J. 2007. "Changes in women's mate preferences across the ovulatory cycle," *Journal of Personality and Social Psychology* 92: 151–63; Grammer, K. 1993. "5-α-androst-16en-3α-on: A male pheromone? A brief report," *Ethology and Sociobiology* 14: 201–8; Havlicek, J., Roberts, S. C., and Flegr, J. 2005. "Women's preference for dominant male odour: Effects of menstrual cycle and relationship status," *Biology Letters* 1: 256–59; Hummel, T., Gollisch, R., Wildt, G., and Kobal, G. 1991. "Changes in olfactory perception during the menstrual cycle," *Experentia* 47: 712–15; Little, A. C., Jones, B. C., and Burriss, R. P. 2007. "Preferences for masculinity in male bodies change across the menstrual cycle," *Hormones and Behavior* 52: 633–39; Lukaszewski, A. W., and Roney, J. R. 2009. "Estimated hormones predict women's mate preferences for dominant personality traits," *Personality and Individual Differences* 47: 191–96; Provost, M. P., Troje, N. F., and Quinsey, V. L. 2008. "Short-term mating strategies and attraction to masculinity in point-light walkers," *Evolution and Human Behavior* 29: 65–69; Puts, D. A. 2005. "Mating context and menstrual phase affect women's preferences for male voice pitch," *Evolution and Human Behavior* 26: 388–

97; Puts, D. A. 2006. "Cyclic variation in women's preferences for masculine traits: Potential hormonal causes," *Human Nature* 17: 114–27.

30. Bellis, M. A., and Baker, R. R. 1990. "Do females promote sperm competition? Data for humans," *Animal Behaviour* 40: 997–99; Gangestad, S. W., Thornhill, R., and Garver, C. E. 2002. "Changes in women's sexual interests and their partners' mate-retention tactics across the menstrual cycle: Evidence for shifting conflicts of interest," *Proceedings of the Royal Society of London B* 269: 975–82; Gangestad, S. W., Thornhill, R., and Garver-Apgar, C. E. 2005. "Women's sexual interests across the ovulatory cycle depend on primary partner developmental instability," *Proceedings of the Royal Society of London B* 272: 2023–27; Haselton, M. G., and Gangestad, S. W. 2006. "Conditional expression of women's desires and men's mate guarding across the ovulatory cycle," *Hormones and Behavior* 49: 509–18; Jones, B. C., Little, A. C., Boothroyd, L. G., DeBruine, L. M., Feinberg, D. R., Law Smith, M. J., Moore, F. R., and Perrett, D. I. 2005. "Commitment to relationships and preferences for femininity and apparent health in faces are strongest on days of the menstrual cycle when progesterone level is high," *Hormones and Behavior* 48: 283–90; Pillsworth, E., and Haselton, M. 2006. "Male sexual attractiveness predicts differential ovulatory shifts in female extra-pair attraction and male mate retention," *Evolution and Human Behavior* 27: 247–58; Guéguen, N. 2009a. "The receptivity of women to courtship solicitation across the menstrual cycle: A field experiment," *Biological Psychology* 80: 321–24; Guéguen, N. 2009b. "Menstrual cycle phases and female receptivity to a courtship solicitation: An evaluation in a nightclub," *Evolution and Human Behavior* 30: 351–55; Durante, K. M., Griskevicius, V., Hill, S. E., Perilloux, C., and Li, N. P. 2011. "Ovulation, female competition, and product choice: Hormonal influences on consumer behavior," *Journal of Consumer Research* 37: 921–35; Durante, K. M., Li, N. P., and Haselton, M. G. 2008. "Changes in women's choice of dress across the ovulatory cycle: Naturalistic and laboratory task-based evidence," *Personality and Social Psychology Bulletin* 34: 1451–60; Haselton, M. G., Mortezaie, M., Pillsworth, E. G., Bleske-Rechek, A., and Frederick, D. A. 2007. "Ovulatory shifts in human female ornamentation: Near ovulation, women dress to impress," *Hormones and Behavior* 51: 40–45; Hill, S. E., and Durante, K. M. 2009. "Do women feel worse to look their best? Testing the relationship between self-esteem and fertility status across the menstrual cycle," *Personality and Social Psychology Bulletin* 35: 1592–601.
31. Gangestad, S. W., Thornhill, R., and Garver-Apgar, C. E. 2005. "Women's sexual interests across the ovulatory cycle depend on primary partner developmental instability," *Proceedings of the Royal Society of London B* 272: 2023–27; Haselton, M. G., and Gangestad, S. W. 2006. "Conditional expression of women's desires and men's mate guarding across the ovulatory cycle," *Hormones and Behavior* 49: 509–18; Pillsworth, E., and Haselton, M. 2006. "Male sexual attractiveness predicts differential ovulatory shifts in female extra-pair attraction and male mate retention," *Evolution and Human Behavior* 27: 247–58. MHC genes: Garver-Apgar, C. E., Gangestad, S. W., Thornhill, R., Miller, R. D., and Olp, J. J. 2006. "Major histocompatibility complex alleles, sexual responsivity, and unfaithfulness in romantic couples," *Psychological Science* 17: 830–35.

32. Bradley, M. M., Miccoli, L., Escrig, M. A., and Lang, P. J. 2008. "The pupil as a measure of emotional arousal and autonomic activation," *Psychophysiology* 45: 602–7; Steinhauer, S. R., Siegle, G. S., Condray, R., and Pless, M. 2004. "Sympathetic and parasympathetic innervation of pupillary dilation during sustained processing," *International Journal of Psychophysiology* 52: 77–86.
33. Van Gerven, P. W. M., Paas, F., Van Merriënboer, J. J. G., and Schmidt, H. G. 2004. "Memory load and the cognitive pupillary response in aging," *Psychophysiology* 41(2): 167–74; Morris, S. K., Granholm, E., Sarkin, A. J., and Jeste, D. V. 1997. "Effects of schizophrenia and aging on pupillographic measures of working memory," *Schizophrenia Research* 27: 119–28; Winn, B., Whitaker, D., Elliott, D. B., and Phillips, N. J. 1994. "Factors affecting light-adapted pupil size in normal human subjects," *Investigative Ophthalmology & Visual Science* (March 1994) 35: 1132–37.
34. Tombs, S., and Silverman, I. 2004. "Pupillometry: A sexual selection approach," *Evolution and Human Behavior* 25: 221–28.
35. Wiseman, R., and Watt, C. 2010. "Judging a book by its cover: The unconscious influence of pupil size on consumer choice," *Perception* 39: 1417–19.
36. Laeng, B., and Falkenberg, L. 2007. "Women's pupillary responses to sexually significant others during the hormonal cycle," *Hormones and Behavior* 52: 520–30.
37. Sammaknejad, N. 2012. "Facial attractiveness: The role of iris size, pupil size, and scleral color." PhD diss., University of California–Irvine.
38. Caryl, P. G., Bean, J. E., Smallwood, E. B., Barron, J. C., Tully, L., and Allerhand, M. 2008. "Women's preference for male pupil-size: Effects of conception risk, sociosexuality and relationship status," *Personality and Individual Differences* 46: 503–8.
39. Ibid.
40. Kobayashi, H., and Kohshima, S. 2001. "Unique morphology of the human eye and its adaptive meaning: Comparative studies on external morphology of the primate eye," *Journal of Human Evolution* 40: 419–35; Hinde, R. A., and Rowell, T. E. 1962. "Communication by posture and facial expression in the rhesus monkey," *Proceedings of the Zoological Society of London* 138: 1–21.
41. Provine, R. R., Cabrera, M. O., Brocato, N. W., and Krosnowski, K. A. 2011. "When the whites of the eyes are red: A uniquely human cue," *Ethology* 117: 1–5.
42. Gründl, M., Knoll, S., Eisenmann-Klein, M., and Prantl, L. 2012. "The blue-eyes stereotype: Do eye color, pupil diameter, and scleral color affect attractiveness?" *Aesthetic Plastic Surgery* 36: 234–40; Provine, R. R., Cabrera, M. O., and Nave-Blodgett, J. 2013. "Red, yellow, and super-white sclera: Uniquely human cues for healthiness, attractiveness, and age," *Human Nature* 24: 126–36.
43. Watson, P. G., and Young, R. D. 2004. "Scleral structure, organization and disease. A review," *Experimental Eye Research* 78: 609–23.
44. Sammaknejad, N. 2012. "Facial attractiveness: The role of iris size, pupil size, and scleral color." PhD diss., University of California–Irvine.
45. Goto, E. 2006. "The brilliant beauty of the eye: Light reflex from the cornea and tear film," *Cornea* 25 (Suppl 1): S78–81; Goto, E., Dogru, M., Sato, E. A., Matsumoto, Y., Takano, Y., and Tsubota, K. 2011. "The sparkle of the eye: The impact of ocular surface wetness on corneal light reflection," *American Journal of Ophthalmology* 151: 691–96; Korb, D. R., Craig, J. P., Doughty, M., Guillon, J. P., Smith, G., and

Tomlinson, A. 2002. *The Tear Film: Structure, Function and Clinical Examination* (Oxford, UK: Butterworth-Heinemann).

46. Ibid.
47. Breakfield, M. P., Gates, J., Keys, D., Kesbeke, F., Wijngaarden, J. P., Monteiro, A., French, V., and Carroll, S. B. 1996. "Development, plasticity and evolution of butterfly eyespot patterns," *Nature* 384: 236–42; French, V., and Breakfield, P. M. 1992. "The development of eyespot patterns on butterfly wings: Morphogen sources or sinks?" *Development* 116: 103–9; Keys, D. N., Lewis, D. L., Selegue, J. E., Pearson, B. J., Goodrich, L. V., Johnson R. L., Gates, J., Scott, M. P., and Carroll, S. B. 1999. "Recruitment of a hedgehog regulatory circuit in butterfly eyespot evolution," *Science* 283: 532–34; Monteiro, A. 2015. "Origin, development, and evolution of butterfly eyespots," *Annual Review of Entomology* 60: 253–71; Reed, R. D., and Serfas, M. S. 2004. "Butterfly wing pattern evolution is associated with changes in a Notch/Distal-less temporal pattern formation process," *Current Biology* 14: 1159–66.
48. Costanzo, K., and Monteiro, A. 2007. "The use of chemical and visual cues in female choice in the butterfly *Bicyclus anynana*," *Proceedings of the Royal Society B* 274: 845–51; Robertson, K. A., and Monteiro, A. 2005. "Female *Bicyclus anynana* butterflies choose males on the basis of their dorsal UV-reflective eyespot pupils," *Proceedings of the Royal Society B* 272: 1541–46.
49. Zahavi, A. 1975. "Mate selection—A selection for a handicap," *Journal of Theoretical Biology* 53(1): 205–14; Zahavi, A., and Zahavi, A. 1997. *The Handicap Principle: A Missing Piece of Darwin's Puzzle* (Oxford, UK: Oxford University Press); Koch, N. 2011. "A mathematical analysis of the evolution of human mate choice traits: Implications for evolutionary psychologists," *Journal of Evolutionary Psychology* 9(3): 219–47.
50. Hamilton, W. 1964. "The genetical evolution of social behaviour. I," *Journal of Theoretical Biology* 7(1): 1–16; Marshall, J. A. R. 2015. *Social Evolution and Inclusive Fitness Theory: An Introduction* (Princeton, NJ: Princeton University Press). For a critique of inclusive fitness, see Nowak, M. A., Tarnita, C. E., and Wilson, E. O. 2010. "The evolution of eusociality," *Nature* 466: 1057–62; Wilson, E. O. 2012. *The Social Conquest of Earth*. New York: Liveright.
51. Mateo, J. M. 1996. "The development of alarm-call response behavior in free-living juvenile Belding's ground squirrels," *Animal Behaviour* 52: 489–505.
52. Dawkins, R. 1979. "12 Misunderstandings of kin selection," *Zeitschrift für Tierpsychologie* 51: 184–200; Park, J. H. 2007. "Persistent misunderstandings of inclusive fitness and kin selection: Their ubiquitous appearance in social psychology textbooks," *Evolutionary Psychology* 5(4): 860–73; West, S. A., Mouden, C. E., and Gardner, A. 2011. "Sixteen common misconceptions about the evolution of cooperation in humans," *Evolution and Social Behaviour* 32: 231–62.
53. Holekamp, K. E. 1986. "Proximal causes of natal dispersal in Belding's ground squirrels," *Ecological Monographs* 56(4): 365–91; Sherman, P. W. 1981. "Kinship, demography, and Belding's ground squirrel nepotism," *Behavioral Ecology and Sociobiology* 8: 251–59.
54. Dal Martello, M. F., and Maloney, L. T. 2010. "Lateralization of kin recognition

signals in the human face," *Journal of Vision* 10(8):9 1–10; Dal Martello, M. F., DeBruine, L. M., and Maloney, L. T. 2015. "Allocentric kin recognition is not affected by facial inversion." *Journal of Vision* 15(13):5 1–11; Maloney, L. T., and Dal Martello, M. F. 2006. "Kin recognition and the perceived facial similarity of children," *Journal of Vision* 6(10): 1047–56.

55. Buss, D. M. 2016. *Evolutionary Psychology: The New Science of the Mind* (New York: Routledge); Etcoff, N. 1999. *Survival of the Prettiest: The Science of Beauty* (New York: Anchor Books, Random House); Perrett, D. 2010. *In Your Face: The New Science of Human Attraction* (New York: Palgrave McMillan). For an argument that our ratings of facial attractiveness are not due to genes but to differences in environment that vary from person to person, see Germine, L., Russell, R., Bronstad, P. M., Blokland, G. A. M., Smoller, J. W., Kwok, H., Anthony, S. E., Nakayama, K., Rhodes, G., and Wilmer, J. B. 2015. "Individual aesthetic preferences for faces are shaped mostly by environments, not genes," *Current Biology* 25: 2684–89.

Chapter Three: Reality

1. Hoffman, D. D. 1998. *Visual Intelligence: How We Create What We See* (New York: W. W. Norton); Knill, D. C., and Richards W. A., eds. 1996. *Perception as Bayesian Inference* (Cambridge, UK: Cambridge University Press); Palmer, S. 1999. *Vision Science: Photons to Phenomenology* (Cambridge, MA: MIT Press); Pinker, S. 1997. *How the Mind Works* (New York: W. W. Norton).
2. Geisler, W. S., and Diehl, R. L. 2002. "Bayesian natural selection and the evolution of perceptual systems," *Philosophical Transactions of the Royal Society of London B* 357: 419–48.
3. Geisler, W. S., and Diehl, R. L. 2003. "A Bayesian approach to the evolution of perceptual and cognitive systems," *Cognitive Science* 27: 379–402.
4. Trivers, R. L. 2011. *The Folly of Fools: The Logic of Deceit and Self-Deception in Human Life* (New York: Basic Books).
5. Noë, A., and O'Regan, J. K. 2002. "On the brain-basis of visual consciousness: A sensorimotor account," in A. Noë and E. Thompson, eds., *Vision and Mind: Selected Readings in the Philosophy of Perception* (Cambridge, MA: MIT Press), 567–98; O'Regan, J. K., and Noë, A. 2001. "A sensorimotor account of vision and visual consciousness," *Behavioral and Brain Sciences* 24: 939–1031. Their ideas are similar to those of Gibson, who argued that we directly perceive, without computations, aspects of the environment critical to survival, such as "affordances"—all the action possibilities in the environment. Gibson, J. J. 1950. *The Perception of the Visual World* (Boston: Houghton Mifflin); Gibson, J. J. 1960. *The Concept of the Stimulus in Psychology, The American Psychologist* 15/1960, 694–703; Gibson, J. J. 1966. *The Senses Considered as Perceptual Systems* (Boston: Houghton Mifflin); Gibson, J. J. 1979. *The Ecological Approach to Visual Perception* (Boston: Houghton Mifflin).
6. Pizlo, Z., Li, Y., Sawada, T., and Steinman, R. M. 2014. *Making a Machine That Sees Like Us* (New York: Oxford University Press).
7. Loomis, J. M., Da Silva, J. A., Fujita, N., and Fukusima, S. S. 1992. "Visual space per-

ception and visually directed action," *Journal of Experimental Psychology: Human Perception and Performance* 18: 906–21; Loomis, J. M., and Philbeck, J. W. 1999. "Is the anisotropy of 3-D shape invariant across scale?" *Perception & Psychophysics* 61: 397–402; Loomis, J. M. 2014. "Three theories for reconciling the linearity of egocentric distance perception with distortion of shape on the ground plane," *Psychology & Neuroscience* 7: 245–51; Foley, J. M., Ribeiro-Filho, N. P., and Da Silva, J. A. 2004. "Visual perception of extent and the geometry of visual space," *Vision Research* 44: 147–56; Wu, B., Ooi, T. L., and He, Z. J. 2004. "Perceiving distance accurately by a directional process of integrating ground information," *Nature* 428: 73–77; Howe, C. Q., and Purves, D. 2002. "Range image statistics can explain the anomalous perception of length," *Proceedings of the National Academy of Sciences* 99: 13184–88; Burge, J., Fowlkes, C. C., and Banks, M. S. 2010. "Natural-scene statistics predict how the figure-ground cue of convexity affects human depth perception," *The Journal of Neuroscience* 30(21): 7269–80; Froyen, V., Feldman, J., and Singh, M. 2013. "Rotating columns: Relating structure-from-motion, accretion/deletion, and figure/ground," *Journal of Vision* 13, doi: 10.1167/13.10.6.

8. Marr, D. 1982. *Visio.* (San Francisco: Freeman Press).
9. Ibid.
10. Pinker, S. 1997. *How the Mind Works* (New York: W. W. Norton).
11. Fodor, J. 2000. *The Mind Doesn't Work That Way* (Cambridge, MA: MIT Press).
12. Pinker, S. 2005. "So how *does* the mind work?" *Mind & Language* 20: 1–24.
13. Ibid.
14. Hawking, S., and Mlodinow, L. 2012. *The Grand Design* (New York: Bantam).
15. Ibid.

Chapter Four: Sensory

1. A 1954 letter of Pauli to Einstein, in Born, M. 1971. *The Born-Einstein Letters* (New York: Walker).
2. Bell, J. S. 1964. "On the Einstein Podolsky Rosen paradox," *Physics* 1: 195–200.
3. Wilkins, J. S., and Griffiths, P. E. 2012. "Evolutionary debunking arguments in three domains: Fact, value, and religion," in J. Maclaurin and G. Dawes, eds., *A New Science of Religion* (New York: Routledge).
4. Darwin, C. 1859. *On the Origin of Species by Means of Natural Selection, or the Preservation of Favoured Races in the Struggle for Life* (London: John Murray, 127).
5. Darwin, C. 1871. *The Descent of Man, and Selection in Relation to Sex* (London: John Murray, 62).
6. Huxley, T. H. 1880. "The coming of age of 'The origin of species,'" *Science* 1: 15–17.
7. Dawkins, R. 1976. *The Selfish Gene* (New York: Oxford University Press).
8. Smolin, L. 1992. "Did the universe evolve?" *Classical and Quantum Gravity* 9: 173–91; Smolin, L. 1997. *The Life of the Cosmos* (Oxford, UK: Oxford University Press).
9. Dawkins, R. 1983. "Universal Darwinism," in D. S. Bendall, ed., *Evolution from Molecules to Man* (Cambridge, UK: Cambridge University Press); Dennett, D. 1996.

Darwin's Dangerous Idea: Evolution and the Meanings of Life (New York: Simon & Schuster).
10. Dennett, D. 1996. *Darwin's Dangerous Idea: Evolution and the Meanings of Life* (New York: Simon & Schuster).
11. Smith, J. M., and Price, G. R. 1973. "The logic of animal conflict," *Nature* 246: 15–18; Nowak, M. A. 2006. *Evolutionary Dynamics: Exploring the Equations of Life* (Cambridge, MA: Belknap Press).
12. Polis, G. A., and Farley, R. D. 1979. "Behavior and ecology of mating in the cannibalistic scorpion *Paruroctonus mesaensis* Stahnke (Scorpionida: Vaejovidae)," *Journal of Arachnology* 7: 33–46.
13. Smith, J. M., and Price, G. R. 1973. "The logic of animal conflict," *Nature* 246: 15–18; Smith, J. M. 1974. "The theory of games and the evolution of animal conflicts," *Journal of Theoretical Biology* 47: 209–21.
14. Ibid.
15. Ibid.
16. Nowak, M. A. 2006. *Evolutionary Dynamics: Exploring the Equations of Life* (Cambridge, MA: Belknap Press).
17. Ibid.
18. Prakash, C., Stephens, K., Hoffman, D. D., and Singh, M. 2017. "Fitness beats truth in the evolution of perception," http://cogsci.uci.edu/~ddhoff/FBT-7-30-17.
19. Mark, J. T., Marion, B., and Hoffman, D. D. 2010. "Natural selection and veridical perceptions," *Journal of Theoretical Biology* 266: 504–15; Marion, B. B. 2013. "The impact of utility on the evolution of perceptions," PhD diss., University of California–Irvine; Mark, J. T. 2013. "Evolutionary pressures on veridical perception: When does natural selection favor truth?" PhD diss., University of California–Irvine.
20. Marr, D. 1982. *Vision* (San Francisco: Freeman Press).
21. Ibid.
22. Hood, B. 2014. *The Domesticated Brain* (London: Penguin); Bailey, D. H., and Geary, D. C. 2009. "Hominid brain evolution: Testing climatic, ecological, and social competition models," *Human Nature* 20: 67–79.
23. Nowak, M. A. 2006. *Evolutionary Dynamics: Exploring the Equations of Life* (Cambridge, MA: Belknap Press).
24. Mark, J. T. 2013. "Evolutionary pressures on veridical perception: When does natural selection favor truth?" PhD diss., University of California–Irvine; Hoffman, D. D., Singh, M., and Mark, J. T. 2013. "Does evolution favor true perceptions?" *Proceedings of the SPIE 8651, Human Vision and Electronic Imaging XVIII*, 865104, doi: 10.1117/12.2011609.
25. Hoffman, D. D., Singh, M., and Prakash, C. 2015. "The interface theory of perception," *Psychonomic Bulletin and Review* 22: 1480–1506.
26. The straw man fallacy is an informal fallacy: one claims to refute an opponent's argument by refuting an argument the opponent did not make.
27. Webster, M. A. 2014. "Probing the functions of contextual modulation by adapting images rather than observers," *Vision Research* 104: 68–79; Webster, M. A. 2015. "Visual adaptation," *Annual Reviews of Vision Science* 1: 547–67.

28. Marion, B. B. 2013. "The impact of utility on the evolution of perceptions," PhD diss., University of California–Irvine.
29. Mausfeld, R. 2015. "Notions such as 'truth' or 'correspondence to the objective world' play no role in explanatory accounts of perception," *Psychonomic Bulletin & Review* 6: 1535–40.
30. Duret, L. 2008. "Neutral theory: The null hypothesis of molecular evolution," *Nature Education* 1(1): 218.
31. Cohen, J. 2015. "Perceptual representation, veridicality, and the interface theory of perception," *Psychonomic Bulletin & Review* 6: 1512–18.
32. Ibid.
33. Cover, T. M., and Thomas, J. A. 2006. *Elements of Information Theory* (Hoboken, NJ: Wiley).
34. For more on the philosophy of perceptual content, see Hawley, K., and Macpherson, F., eds. 2011. *The Admissible Contents of Experience* (West Sussex, UK: Wiley-Blackwell); Siegel, S. 2011. *The Contents of Visual Experience* (Oxford, UK: Oxford University Press); Brogard, B., ed. 2014. *Does Perception Have Content*? (Oxford, UK: Oxford University Press).
35. Foreword to Dawkins, R. 1976. *The Selfish Gene* (New York: Oxford University Press).
36. Pinker, S. 1997. *How the Mind Works* (New York: W. W. Norton).

Chapter Five: Illusory

1. Hoffman, D. D. 1998. *Visual Intelligence: How We Create What We See* (New York: W. W. Norton); Hoffman, D. D. 2009. "The interface theory of perception," in S. Dickinson, M. Tarr, A. Leonardis, and B. Schiele, eds., *Object Categorization: Computer and Human Vision Perspectives* (New York: Cambridge University Press), 148–65; Hoffman, D. D. 2011. "The construction of visual reality," in J. Blom and I. Sommer, eds., *Hallucinations: Theory and Practice* (New York: Springer, 7–15); Hoffman, D. D. 2012. "The sensory desktop," in J. Brockman, ed., *This Will Make You Smarter: New Scientific Concepts to Improve Your Thinking* (New York: Harper Perennial), 135–38; Hoffman, D. D. 2013. "Public objects and private qualia: The scope and limits of psychophysics," in L. Albertazzi, ed., *The Wiley-Blackwell Handbook of Experimental Phenomenology* (New York: Wiley-Blackwell), 71–89; Hoffman, D. D. 2016. "The interface theory of perception," *Current Directions in Psychological Science* 25(3): 157–61; Hoffman, D. D. 2018. "The interface theory of perception," in *Stevens' Handbook of Experimental Psychology and Cognitive Neuroscience,* 4th edition (Hoboken, NJ: Wiley); Hoffman, D. D., and Prakash, C. 2014. "Objects of consciousness," *Frontiers in Psychology: Perception Science*, http://dx.doi.org/10.3389/fpsyg.2014.00577; Hoffman, D. D., Singh, M., and Prakash, C. 2015. "The interface theory of perception," *Psychonomic Bulletin and Review* 22: 1480–1506; Hoffman, D. D., Singh, M., and Mark, J. T. 2013. "Does evolution favor true perceptions?" *Proceedings of the SPIE 8651, Human Vision and Electronic Imaging XVIII*, 865104, doi: 10.1117/12.2011609; Koenderink, J. J. 2011. "Vision as a user interface," *Human Vision and Electronic Imaging XVI, SPIE* Vol. 7865, doi: 10.1117/12.881671; Koenderink, J. J. 2013. "World, envi-

ronment, umwelt, and inner-world: A biological perspective on visual awareness," *Human Vision and Electronic Imaging XVIII, SPIE* Vol. 8651, doi: 10.1117/12.2011874; Mark, J. T., Marion, B., and Hoffman, D. D. 2010. "Natural selection and veridical perceptions," *Journal of Theoretical Biology* 266: 504–15; Mausfeld, R. 2002. "The physicalist trap in perception theory," in D. Heyer and R. Mausfeld, eds., *Perception and the Physical World: Psychological and Philosophical Issues in Perception* (New York: Wiley), 75–112; Singh, M., and Hoffman, D. D. 2013. "Natural selection and shape perception: Shape as an effective code for fitness," in S. Dickinson and Z. Pizlo, eds., *Shape Perception in Human and Computer Vision: An Interdisciplinary Perspective* (New York: Springer), 171–85. For the related idea of an Umwelt, see von Uexküll, J. 1909. *Umwelt und Innenwelt der Tiere* (Berlin: Springer-Verlag); von Uexküll, J. 1926. *Theoretical Biology* (New York: Harcourt, Brace); von Uexküll, J. 1957. "A stroll through the worlds of animals and men: A picture book of invisible worlds," in C. H. Schiller, ed., *Instinctive Behavior: Development of a Modern Concept* (New York: Hallmark); Boyer, P. 2001. "Natural epistemology or evolved metaphysics? Developmental evidence for early-developed, intuitive, category-specific, incomplete, and stubborn metaphysical presumptions," *Philosophical Psychology* 13: 277–97.

2. Shermer, M. 2015. "Did humans evolve to see things as they really are? Do we perceive reality as it is?" *Scientific American* (November), https://www.scientificamerican.com/article/did-humans-evolve-to-see-things-as-they-really-are/.
3. Berkeley, G. 1710. *A Treatise Concerning the Principles of Human Knowledge.*
4. Kant, I. 1781. *Critique of Pure Reason* (New York: American Home Library).
5. Stroud, B. 1999. *The Quest for Reality: Subjectivism and the Metaphysics of Color* (Oxford, UK: Oxford University Press).
6. Strawson, P. F. 1990. *The Bounds of Sense: An Essay on Kant's Critique of Pure Reason* (London: Routledge), 38.
7. von Uexküll, J. 1934. *A Foray into the Worlds of Animals and Humans* (Berlin: Springer).
8. Plato, *Republic.*
9. Palmer, S. 1999. *Vision Science: Photons to Phenomenology* (Cambridge, MA: MIT Press).
10. See, e.g., Plantinga, A. 2011. *Where the Conflict Really Lies: Science, Religion and Naturalism* (New York: Oxford University Press); Balfour, A. J. 1915. *Theism and Humanism, Being the Gifford Lectures Delivered at the University of Glasgow, 1914* (New York: Hodder & Stoughton).
11. Cosmides, L., and Tooby, J. 1992. "Cognitive Adaptions for Social Exchange," in Barkow, J., Cosmides, L., and Tooby, J., eds., *The adapted mind: Evolutionary psychology and the generation of culture* (New York: Oxford University Press).
12. Mercier, H., and Sperber, D. 2011. "Why do humans reason? Arguments for an argumentative theory," *Behavioral and Brain Sciences* 34: 57–111; Mercier, H., and Sperber, D. 2017. *The Enigma of Reason* (Cambridge, MA: Harvard University Press).
13. Shermer, M. 2015. "Did humans evolve to see things as they really are? Do we perceive reality as it is?" *Scientific American* (November), https://www.scientificamerican.com/article/did-humans-evolve-to-see-things-as-they-really-are/.

14. There is a technical issue here about fitness payoffs. In this chapter I say that it is helpful to distinguish two different senses of real: existing, and existing when unperceived. The latter sense of real I call objective reality, and I argue that our senses evolved to track fitness payoffs rather than objective reality. But fitness payoffs, as a mathematical abstraction, might exist when unperceived. Suppose, for instance, that I am in a deep, dreamless sleep, and that therefore I am, arguably, not perceiving anything. Nevertheless, it seems plausible to claim that my fitness payoffs do still exist, even though I am not perceiving them. After all, my fitness might decrease if, for instance, I fell out of bed while sound asleep. So, my fitness payoffs are objective; they exist when unperceived. Fair enough. But my fitness payoffs would not exist if I did not exist. There is a stronger sense of objective, let's call it "strongly objective," in which something is real if it exists even if no perceiver exists. Many physicists, for instance, claim that spacetime and objects existed before there were any organisms to perceive them, and that therefore spacetime and objects are strongly objective. Fitness payoffs, however, do not exist unless there are organisms, and thus are not strongly objective. When I speak of evolution shaping organisms whose perceptions track fitness rather than truth, the "truth" that I have in mind is the physicists' notion of a strongly objective reality.

Chapter Six: Gravity

1. A 1954 letter of Pauli to Einstein, in Born, M. 1971. *The Born-Einstein Letters* (New York: Walker).
2. A 1948 letter of Einstein to Born, in Born, M. 1971. *The Born-Einstein Letters* (New York: Walker).
3. Ibid.
4. Bell, J. S. 1964. "On the Einstein Podolsky Rosen paradox," *Physics* 1: 195–200.
5. Hensen, B., et al. 2015. "Loophole-free Bell inequality violation using electron spins separated by 1.3 kilometres," *Nature* 526: 682–86.
6. Ibid.
7. Giustina, M., et al. 2015. "Significant-loophole-free test of Bell's Theorem with entangled photons," *Physical Review Letters* 115: 250401; Gröblacher, S., Paterek, T., Kaltenbaek, R., Brukner, Č., Żukowski, M., Aspelmeyer, M., and Zeilinger, A. 2007. "An experimental test of non-local realism," *Nature* 446: 871–75.
8. Gröblacher, S., Paterek, T., Kaltenbaek, R., Brukner, Č., Żukowski, M., Aspelmeyer, M., and Zeilinger, A. 2007. "An experimental test of non-local realism," *Nature* 446: 871–75.
9. Bell, J. S. 1966. "On the problem of hidden variables in quantum mechanics," *Reviews of Modern Physics* 38: 447–52; Kochen, S., and Specker, E. P. 1967. "The problem of hidden variables in quantum mechanics," *Journal of Mathematics and Mechanics* 17: 59–87. For a wide-ranging discussion of contextuality, see Dzhafarov, E., Jordan, S., Zhang, R., and Cervantes, V., eds. 2016. *Contextuality from quantum physics to psychology* (Singapore: World Scientific).

10. Einstein, A., Podolsky, B., and Rosen, N. 1935. "Can quantum-mechanical description of physical reality be considered complete?" *Physical Review* 47: 777–80.
11. Cabello, A., Estebaranz, J. M., and García-Alcaine, G. 1996. "Bell-Kochen-Specker Theorem: A proof with 18 vectors," *Physics Letters A* 212: 183. See also Klyachko, A. A., Can, M. A., Binicioglu, S. and Shumovsky, A. S. 2008. "Simple test for hidden variables in spin-1 systems," *Physical Review Letters* 101: 020403.
12. Formaggio, J. A., Kaiser, D. I., Murskyj, M. M., and Weiss, T. E. 2016. "Violation of the Leggett-Garg inequality in neutrino oscillations," arXiv:1602.00041 [quant-ph].
13. Rovelli, C. 1996. "Relational quantum mechanics," *International Journal of Theoretical Physics* 35: 1637–78.
14. Ibid.
15. Ibid.
16. Fields, C. 2016. "Building the observer into the system: Toward a realistic description of human interaction with the world," *Systems* 4: 32, doi: 10.3390/systems4040032.
17. Fuchs, C. A., Mermin, N. D., and Schack, R. 2014. "An introduction to QBism with an application to the locality of quantum mechanics," *American Journal of Physics* 82: 749.
18. Ibid.
19. Fuchs, C. 2010. "QBism, the perimeter of quantum Bayesianism," arXiv:1003.5209 v51. See also the summary of QBism in von Baeyer, H. C. 2016. *QBism: The Future of Quantum Physics* (Cambridge, MA: Harvard University Press), and the critique of QBism in Fields, C. 2012. "Autonomy all the way down: Systems and dynamics in quantum Bayesianism," arXiv:1108.2024v2 [quant-ph].
20. Bartley, W. W. 1987. "Philosophy of biology versus philosophy of physics," in G. Radnitzky and W. W. Bartley III, eds., *Evolutionary Epistemology, Theory of Rationality, and the Sociology of Knowledge* (La Salle, IL: Open Court).
21. Ibid.
22. Wheeler, J. A. 1979. "Beyond the black hole," in H. Woolfe, ed., *Some Strangeness in the Proportion: A Centennial Symposium to Celebrate the Achievements of Albert Einstein* (Reading, PA: Addison-Wesley), 341–75.
23. Wheeler, J. A. 1978. "The 'past' and the 'delayed-choice' double-slit experiment," in A. R. Marlow, ed., *Mathematical Foundations of Quantum Theory* (New York: Academic).
24. Ibid.
25. Eibenberger, S., Gerlich, S., Arndt, M., Mayor, M., and Tüxen, J. 2013. "Matter–wave interference of particles selected from a molecular library with masses exceeding 10,000 amu," *Physical Chemistry Chemical Physics* 15: 14696.
26. Wheeler, J. A. 1979. "Beyond the black hole," in H. Woolfe, ed., *Some Strangeness in the Proportion: A Centennial Symposium to Celebrate the Achievements of Albert Einstein* (Reading, PA: Addison-Wesley), 341–75.
27. Jacques, V., Wu, E., Grosshans, F., Treussart, F., Grangier, P., Aspect, A., and Roch, J-F. 2007. "Experimental realization of Wheeler's delayed-choice gedanken experiments," *Science* 315(5814): 966–68; Manning, A. G., Khakimov, R. I., Dall, R. G.,

and Truscott, A. G. 2015. "Wheeler's delayed-choice gedanken experiment with a single atom," *Nature Physics* 11: 539–42.

28. Ibid.
29. Wheeler, J. A. 1990. "Information, physics, quantum: The search for links," in W. H. Zurek, ed., *Complexity, Entropy, and the Physics of Information, SFI Studies in the Sciences of Complexity,* vol. VIII (New York: Addison-Wesley).
30. Ibid.
31. Ibid.
32. Bekenstein, J. D. 1981. "Universal upper bound on the entropy-to-energy ratio for bounded systems," *Physical Review D* 23: 287–98; Bekenstein, J. D. 2003. "Information in the Holographic Universe: Theoretical results about black holes suggest that the universe could be like a gigantic hologram," *Scientific American* (August), 59; Susskind, L. 2008. *The Black Hole War* (New York: Little, Brown).
33. This raises the unsolved "Lorentz invariance violation" problem in physics.
34. Susskind, L. 2008. *The Black Hole War* (New York: Little, Brown).
35. Ibid.
36. Quantum information theory differs from classical information theory because, as Fuchs (2010) puts it, "quantum mechanics is an addition to Bayesian probability theory—not a generalization of it, not something orthogonal to it altogether, but an addition." In particular, the Born Rule is "a functional of a usage of the Law of Total Probability that one would have made in another (counterfactual) context." Fuchs, C. 2010. "QBism, the perimeter of quantum Bayesianism," arXiv:1003.5209v51. See also D'Ariano, G. M., Chiribella, G., and Perinotti, P. 2017. *Quantum Theory from First Principles: An Informational Approach* (New York: Cambridge University Press).
37. Susskind, L. 2008. *The Black Hole War* (New York: Little, Brown).
38. Ibid.
39. Almheiri, A., Marolf, D., Polchinski, J., and Sully, J. 2013. "Black holes: complementarity or firewalls?" *Journal of High Energy Physics* 2, arXiv:1207.3123.
40. Harlow, D., and Hayden, P. 2013. "Quantum computation vs. firewalls," *Journal of High Energy Physics* 85, https://arxiv.org/abs/1301.4504.
41. Bousso, R. 2012. "Observer complementarity upholds the equivalence principle," arXiv:1207.5192 [hep-th].
42. Gefter, A. 2014. *Trespassing on Einstein's Lawn* (New York: Bantam Books).
43. Fuchs, C. A., Mermin, N. D., and Schack, R. 2014. "An introduction to QBism with an application to the locality of quantum mechanics," *American Journal of Physics* 82: 749.
44. Hawking, S., and Hertog, T. 2006. "Populating the landscape: A top-down approach," *Physical Review D* 73: 123527.
45. Ibid.
46. Ibid.
47. Ibid.
48. Wheeler, J. A. 1982. "Bohr, Einstein, and the strange lesson of the quantum," in R. Q. Elvee, ed., *Mind in Nature: Nobel Conference XVII*, Gustavus Adolphus College, St. Peter, Minnesota (San Francisco: Harper & Row), 1–23.

49. Fuchs, C. 2010. "QBism, the perimeter of quantum Bayesianism," arXiv:1003.5209v51.
50. For an overview of some other interpretations of quantum theory, see, e.g., Albert, D. 1992. *Quantum Mechanics and Experience* (Cambridge, MA: Harvard University Press); Becker, A. 2018. *What Is Real? The Unfinished Quest for the Meaning of Quantum Physics* (New York: Basic Books).
51. https://www.youtube.com/watch?v=U47kyV4TMnE, at 6 minutes, 10 seconds; see also https://www.youtube.com/watch?v=82NatoryBBk&feature=youtu.be.

Chapter Seven: Virtuality

1. Gross, D. 2005. "Einstein and the search for unification," *Current Science* 89: 2035–40.
2. Ibid, 2039.
3. Cole, K. C. 1999. "Time, space obsolete in new view of universe," *Los Angeles Times*, November 16.
4. Singh, M., and Hoffman, D. D. 2013. "Natural selection and shape perception: Shape as an effective code for fitness," in S. Dickinson and Z. Pizlo, eds., *Shape Perception in Human and Computer Vision: An Interdisciplinary Perspective* (New York: Springer), 171–85.
5. Zadra, J. R., Weltman, A. L., and Proffitt, D. R. 2016. "Walkable distances are bioenergetically scaled," *Journal of Experimental Psychology: Human Perception and Performance* 42: 39–51. But such results might be due to optimal coding or to demand characteristics of the experiments. See, e.g., Durgin, F. H., and Li, Z. 2011. "Perceptual scale expansion: An efficient angular coding strategy for locomotor space," *Attention, Perception & Psychophysics* 73: 1856–70.
6. Cover, T. M., and Thomas, J. A. 2006. *Elements of Information Theory* (Hoboken, NJ: Wiley).
7. Almheiri, A., Dong, X., and Harlow, D. 2015. "Bulk locality and quantum error correction in AdS/CFT," arXiv:1411.7041v3 [hep-th].
8. Ibid.
9. Pastawski, F., Yoshida, B., Harlow, B., and Preskill, J. 2015. "Holographic quantum error-correcting codes: Toy models for the bulk/boundary correspondence," arXiv:1503.06237 [hep-th]; Pastawski, F., and Preskill, J. 2015. "Code properties from holographic geometries," arXiv:1612.00017v2 [quant-ph].
10. Pizlo, Z., Li, Y., Sawada, T., and Steinman, R. M. 2014. *Making a Machine That Sees Like Us* (New York: Oxford University Press).
11. Hoffman, D. D., and Prakash, C. 2014. "Objects of consciousness," *Frontiers in Psychology: Perception Science*, http://dx.doi.org/10.3389/fpsyg.2014.00577; see also Terekhov, A. V., and O'Regan, J. K. 2016. "Space as an invention of active agents," *Frontiers in Robotics and AI*, doi: 10.3389/frobt.2016.00004.
12. Symmetries can be described mathematically using group theory. Group theory is a critical tool in the construction of many error-correcting codes. See, e.g., Togneri, R., and deSilva, C. J. S. 2003. *Fundamentals of Information Theory and Coding*

Design (New York: Chapman & Hall/CRC). See also this lecture by Neil Sloane: https://www.youtube.com/watch?v=uCeTOjIlfIg.

13. Pizlo, Z., Li, Y., Sawada, T., and Steinman, R. M. 2014. *Making a Machine That Sees Like Us* (New York: Oxford University Press).
14. Knill, D. C., and Richards W. A., eds. 1996. *Perception as Bayesian Inference* (Cambridge, UK: Cambridge University Press).
15. Varela, F. J., Thompson, E., and Rosch, E. 1991. *The Embodied Mind* (Cambridge, MA: MIT Press).
16. Chemero, A. 2009. *Radical Embodied Cognitive Science* (Cambridge, MA: MIT Press).
17. Rubino, G., Rozema, L. A., Feix, A., Araújo, M., Zeuner, J. M., Procopio, L. M., Brukner, Č., and Walther, P. 2017. "Experimental verification of an indefinite causal order," *Science Advances* 3: e1602589, arXiv:1608.01683v1 [quant-ph].
18. Ibid.
19. Oizumi, M., Albantakis, L., and Tononi, G. 2014. "From the phenomenology to the mechanisms of consciousness: Integrated information theory 3.0," *PLOS Computational Biology* 10: e1003588; Hoel, E. P. 2017. "When the map is better than the territory," *Entropy* 19: 188, doi: 10.3390/e19050188; Searle, J. R. 1998. *Mind, Language and Society: Philosophy in the real world* (New York: Basic Books); Searle, J. R. 2015. *Seeing Things as They Are: A Theory of Perception* (New York: Oxford University Press).
20. Rubino, G., Rozema, L. A., Feix, A., Araújo, M., Zeuner, J. M., Procopio, L. M., Brukner, Č., and Walther, P. 2017. "Experimental verification of an indefinite causal order," *Science Advances* 3: e1602589, arXiv:1608.01683v1 [quant-ph].
21. Cover, T. M., and Thomas, J. A. 2006. *Elements of Information Theory* (Hoboken, NJ: Wiley).
22. Fuchs, C. 2010. "QBism, the perimeter of quantum Bayesianism," arXiv:1003.5209 v51. Fuchs notes that any quantum state written in terms of complex amplitudes can be rewritten with standard probabilities. Quantum theory does not extend standard probability theory, but is simply a model within standard probability theory.
23. A full description of the subjective Necker cube can be found in Bradley, D. R., and Petry, H. M. "Organizational determinants of subjective contour. The subjective Necker cube." *American Journal of Psychology,* 1977, 90, 253–62.
24. Van Raamsdonk, M. 2010. "Building up spacetime with quantum entanglement," *General Relativity and Gravitation* 42: 2323–29; Swingle, B. 2009. "Entanglement renormalization and holography," arXiv:0905.1317 [cond-mat.str-el]; Cao, C., Carroll, S. M., and Michalakis, S. 2017. "Space from Hilbert space: Recovering geometry from bulk entanglement," *Physical Review D* 95: 024031.
25. Morgenstern, Y., Murray, R. F., and Harris, L. R. 2011. "The human visual system's assumption that light comes from above is weak," *Proceedings of the National Academy of Sciences USA* 108(30): 12551–3, doi: 10.1073/pnas.1100794108.
26. For examples of Body Optix™, see http://leejeans-ap.com/bodyoptixdenim/en/index.html and https://www.forbes.com/sites/rachelarthur/2017/09/20/lee-jeans-visual-science-instagram/#220b69987fb2.

Chapter Eight: Polychromy

1. Koenderink, J. 2010. *Color for the Sciences* (Cambridge, MA: MIT Press).
2. Pinna, B., Brelstaff, G., and Spillmann, L. 2001. "Surface color from boundaries: A new 'watercolor' illusion," *Vision Research* 41: 2669–76.
3. van Tuijl, H. F. J. M., and Leeuwenberg, E. L. J. 1979. "Neon color spreading and structural information measures," *Perception & Psychophysics* 25: 269–84; Watanabe, T., and Sato, T. 1989. "Effects of luminance contrast on color spreading and illusory contour in the neon color spreading effect," *Perception & Psychophysics* 45: 427–30.
4. Albert, M., and Hoffman, D. D. 2000. "The generic-viewpoint assumption and illusory contours," *Perception* 29: 303–12; Hoffman, D. D. 1998. *Visual Intelligence: How We Create What We See* (New York: W. W. Norton).
5. The movie is posted at http://www.cogsci.uci.edu/~ddhoff/BB.mp4.
6. Cicerone, C., and Hoffman, D. D. 1997. "Color from motion: Dichoptic activation and a possible role in breaking camouflage," *Perception* 26: 1367–80; Hoffman, D. D. 1998. *Visual Intelligence: How We Create What We See* (New York: W. W. Norton).
7. Labrecque, L. I., and Milne, G. R. 2012. "Exciting red and competent blue: The importance of color in marketing," *Journal of the Academy of Marketing Science* 40: 711–27.
8. Chamovitz, D. 2012. *What a Plant Knows* (New York: Scientific American / Farrar, Straus and Giroux).
9. Ibid.
10. Ibid.
11. Ibid.
12. Wiltbank, L. B., and Kehoe, D. M. 2016. "Two cyanobacterial photoreceptors regulate photosynthetic light harvesting by sensing teal, green, yellow and red light," mBio 7(1): e02130-15, doi: 10.1128/mBio.02130-15.
13. Palmer, S. E., and Schloss, K. B. 2010. "An ecological valence theory of human color preference," *Proceedings of the National Academy of Sciences of the USA* 107: 8877–82; Palmer, S. E., Schloss, K. B., and Sammartino, J. 2013. "Visual aesthetics and human preference," *Annual Review of Psychology* 64: 77–107.
14. I coined the term *chromature* in 2009. It is mentioned in this article in CNN: https://www.cnn.com/2018/04/26/health/colorscope-benefits-of-a-colorful-life/index.html.
15. The number of particles in the observable universe, known as the Eddington number, is roughly 10^{80}, excluding dark matter. If each pixel in an image has 24 bits of color (8 bits each for red, green, and blue), then each pixel has 16,777,216 possible colors. In this case, a patch of pixels has possible chromatures, which dwarfs the Eddington number.
16. Imura, T., Masuda, T., Wada, Y., Tomonaga, M., and Okajima, K. 2016. "Chimpanzees can visually perceive differences in the freshness of foods," *Nature* 6: 34685, doi: 10.1038/srep34685.

17. Cytowic, R. E., and Eagleman, D. M. 2009. *Wednesday Is Indigo Blue: Discovering the Brain of Synesthesia* (Cambridge, MA: MIT Press).
18. Nabokov, V. 1951. *Speak, Memory* (New York: Harper & Bros.).
19. Cytowic, R. E., and Eagleman, D. M. 2009. *Wednesday Is Indigo Blue: Discovering the Brain of Synesthesia* (Cambridge, MA: MIT Press).
20. Cytowic, R. E. 1993. *The Man Who Tasted Shapes* (Cambridge, MA: MIT Press).
21. Cytowic, R. E., and Eagleman, D. M. 2009. *Wednesday Is Indigo Blue: Discovering the Brain of Synesthesia* (Cambridge, MA: MIT Press).
22. Ibid.
23. Asher, Julian E., Lamb, Janine A., Brocklebank, Denise, Cazier, Jean-Baptiste, Maestrini, Elena, Addis, Laura, Sen, Mallika, Baron-Cohen, Simon, and Monaco, Anthony P. 2009. "A whole-genome scan and fine-mapping linkage study of auditory-visual synesthesia reveals evidence of linkage to chromosomes 2q24, 5q33, 6p12, and 12p12," *American Journal of Human Genetics* 84(2): 279–85; Tomson, S. N., Avidan, N., Lee, K., Sarma, A. K., Tushe, R., Milewicz, D. M., Bray, M., Lealc, S. M., and Eagleman, D. M. 2011. "The genetics of colored sequence synesthesia: Suggestive evidence of linkage to 16q and genetic heterogeneity for the condition," *Behavioural Brain Research* 223: 48–52. There may also be important environmental influences on synesthesia. Witthoft and Winawer (2006) report that synesthetic colors might be determined by having colored refrigerator magnets in childhood: Witthoft, N., and Winawer, J. 2006. "Synesthetic colors determined by having colored refrigerator magnets in childhood," *Cortex* 42(2): 175–83.
24. Novich, S. D., Cheng, S., and Eagleman, D. M. 2011. "Is synesthesia one condition or many? A large-scale analysis reveals subgroups," *Journal of Neuropsychology* 5: 353–71.
25. Hubbard, E. M., and Ramachandran, V. S. 2005. "Neurocognitive mechanisms of synesthesia," *Neuron* 48: 509–20; Ramachandran, V. S., and Hubbard, E. M. 2001. "Psychophysical investigations into the neural basis of synaesthesia," *Proceedings of the Royal Society of London B* 268: 979–83.
26. Rouw, R., and Scholte, H. S. 2007. "Increased structural connectivity in grapheme-color synesthesia," *Nature Neuroscience* 10: 792–97.
27. Smilek, Daniel, Dixon, Mike J., Cudahy, Cera, and Merikle, Philip M. 2002. "Synesthetic color experiences influence memory," *Psychological Science* 13(6): 548.
28. Tammet, D. 2006. *Born on a Blue Day* (London: Hodder & Stoughton).
29. Banissy, M. J., Walsh, V., and Ward, J. 2009. "Enhanced sensory perception in synaesthesia," *Experimental Brain Research* 196: 565–71.
30. Havlik, A. M., Carmichael, D. A., and Simner, J. 2015. "Do sequence-space synaesthetes have better spatial imagery skills? Yes, but there are individual differences," *Cognitive Processing* 16(3): 245–53; Simner, J. 2009. "Synaesthetic visuo-spatial forms: Viewing sequences in space," *Cortex* 45: 1138–47; Simner, J., and Hubbard, E. M., eds. 2013. *The Oxford Handbook of Synesthesia* (Oxford, UK: Oxford University Press).
31. Cytowic, R. E. 1993. *The Man Who Tasted Shapes* (Cambridge, MA: MIT Press).
32. Ibid.

33. This example was suggested by Rob Reid.
34. Corcoran, Aaron J., Barber, J. R., and Conner, W. E. 2009. "Tiger moth jams bat sonar," *Science* 325 (5938): 325–27, doi: 10.1126/science.1174096.

Chapter Nine: Scrutiny

1. Tovée, M. J. 2008. *An Introduction to the Visual System* (Cambridge, UK: Cambridge University Press).
2. Li, Z. 2014. *Understanding Vision: Theory, Models, and Data* (Oxford, UK: Oxford University Press).
3. Rensink, R. A., O'Regan, J. K., and Clark, J. J. 1997. "To See or Not to See: The Need for Attention to Perceive Changes in Scenes," *Psychological Science* 8: 368–73.
4. See, e.g., https://www.youtube.com/watch?v=VkrrVozZR2c.
5. Itti, L. 2005. "Quantifying the contribution of low-level saliency to human eye movements in dynamic scenes," *Visual Cognition* 12: 1093–1123; Wolfe, J. M., and Horowitz, T. S. 2004. "What attributes guide the deployment of visual attention and how do they do it?" *Nature Reviews Neuroscience* 5: 495–501.
6. The role of visual attention in marketing is explored in Wedel, M., and Pieters, R., eds. 2008. *Visual Marketing: From Attention to Action* (New York: Lawrence Erlbaum).
7. Li, Z. 2014. *Understanding Vision: Theory, Models, and Data* (Oxford, UK: Oxford University Press); Sprague, T., Itthipuripat, S., and Serences, J. 2018. "Dissociable signatures of visual salience and behavioral relevance across attentional priority maps in human cortex," *Journal of Neurophysiology* http://dx.doi.org/10.1101/196642. I speak here as though neurons exist when not perceived and can perform activities such as signaling. This is just a useful shorthand, using the language of our interface.
8. Navalpakkam, V., and Itti, L. 2007. "Search goal tunes visual features optimally," *Neuron* 53: 605–17.
9. New, J., Cosmides, L., and Tooby, J. 2007. "Category-specific attention for animals reflects ancestral priorities, not expertise," *Proceedings of the National Academy of Sciences* 104: 16598–603.
10. For instance, Paras and Webster (2013) had observers look at images of 1/f noise, and found that two dark spots were sufficient to trigger a face percept, leading the observers to reinterpret the rest of the image as a face. Paras, C., and Webster, M. 2013. "Stimulus requirements for face perception: An analysis based on 'totem poles,'" *Frontiers in Psychology* 4: 18, http://journal.frontiersin.org/article/10.3389/fpsyg.2013.00018/full.
11. Barrett, D. 2010. *Supernormal Stimuli: How Primal Urges Overran Their Evolutionary Purpose* (New York: W. W. Norton).
12. Najemnik, J., and Geisler, W. 2005. "Optimal eye movement strategies in visual search," *Nature* 434: 387–91; Pomplun, M. 2006. "Saccadic selectivity in complex visual search displays," *Vision Research* 46: 1886–1900.
13. Doyle, J. F., and Pazhoohi, F. 2012. "Natural and augmented breasts: Is what is *not* natural most attractive?" *Human Ethology Bulletin* 27: 4.

14. Rhodes, G., Brennan, S., and Carey, S. 1987. "Identification and ratings of caricatures: Implications for mental representations of faces," *Cognitive Psychology* 19(4): 473–97; Benson, P. J., and Perrett, D. I. 1991. "Perception and recognition of photographic quality facial caricatures: Implications for the recognition of natural images," *European Journal of Cognitive Psychology* 3(1): 105–35.
15. Barrett, D. 2010. *Supernormal Stimuli: How Primal Urges Overran Their Evolutionary Purpose* (New York: W. W. Norton).
16. A good example is Etcoff, N., Stock, S., Haley, L. E., Vickery, S. A., and House, D. M. 2011. "Cosmetics as a feature of the extended human phenotype: Modulation of the perception of biologically important facial signals," *PLoS ONE* 6(10): e25656; doi: 10.1371/journal.pone.0025656.
17. Jacobs, G. H. 2009. "Evolution of color vision in mammals," *Philosophical Transactions of the Royal Society B* 364: 2957–67; Melin, A. D., Hiramatsu, C., Parr, N. A., Matsushita, Y., Kawamura, S., and Fedigan, L. M. 2014. "The behavioral ecology of color vision: Considering fruit conspicuity, detection distance and dietary importance," *International Journal of Primatology* 35: 258–87; Hurlbert, A. C., and Ling, Y. 2007. "Biological components of sex differences in color preference," *Current Biology* 17(16): R623–R625.
18. New, J., Krasnow, M. M., Truxaw, D., and Gaulin, S. J. C. 2007. "Spatial adaptations for plant foraging: Women excel and calories count," *Proceedings of the Royal Society, B* 274: 2679–84.
19. Jaeger, S. R., Antúnez, L., Gastón, Aresb, Johnston, J. W., Hall, M., and Harker, F. R. 2016. "Consumers' visual attention to fruit defects and disorders: A case study with apple images," *Postharvest Biology and Technology* 116: 36–44.

Chapter Ten: Community

1. Gross, D. 2005. "Einstein and the search for unification," *Current Science* 89: 2035–40; Cole, K. C. 1999. "Time, space obsolete in new view of universe," *Los Angeles Times*, November 16.
2. Hameroff, S., and Penrose, R. 2014. "Consciousness in the universe: A review of the 'Orch OR' theory," *Physics of Life Reviews* 11: 39–78.
3. Oizumi, M., Albantakis, L., and Tononi, G. 2014. "From the phenomenology to the mechanisms of consciousness: Integrated information theory 3.0," *PLOS Computational Biology* 10: e1003588; see also Hoel, E. P. 2017. "When the map is better than the territory," *Entropy* 19: 188, doi: 10.3390/e19050188.
4. A precise definition of *conscious agent* is given in the Appendix.
5. Pinker, S. 2018. *Enlightenment Now: The Case for Reason, Science, Humanism, and Progress* (New York: Viking).
6. In propositional logic, *modus tollens* is a valid argument form. It says that if P implies Q, and it is not the case that P, then it is not the case that Q. Here is an example: If Pat has lived eighty years, then Pat has lived thirty years. Pat has not lived thirty years. Therefore Pat has not lived eighty years.
7. Einstein, A. 1934. "On the method of theoretical physics," *Philosophy of Science* 1: 163–69.

8. Russell, B. 1924/2010. *The Philosophy of Logical Atomism* (New York: Routledge).
9. A precise definition of *conscious agent* is given in the Appendix.
10. For progress on these issues, see Fields, C., Hoffman, D. D., Prakash, C., and Singh, M. 2017. "Conscious agent networks: Formal analysis and application to cognition," *Cognitive Systems Research* 47: 186–213.
11. For progress on these issues, see Fields, C., Hoffman, D. D., Prakash, C., and Prentner, R. 2017. "Eigenforms, interfaces and holographic encoding: Toward an evolutionary account of objects and spacetime. *Constructivist Foundations* 12(3): 265–74.
12. For an overview of panpsychism, see the article on panpsychism in the online *Stanford Encyclopedia of Philosophy.* It is sometimes claimed that panpsychism is not a dualism. To support this claim, a mathematically precise scientific theory of panpsychism needs to be constructed that is manifestly nondualistic. There is no such theory to date. Integrated information theory (IIT) is often taken to imply panpsychism. According to IIT, "an experience is a maximally irreducible conceptual structure (MICS, a constellation of concepts in qualia space), and the set of elements that generates it constitutes a complex. According to IIT, a MICS specifies the quality of an experience." But, as we have discussed, IIT has not been able to specify a complex for the MICS of even one specific experience, such as the smell of garlic. Until it does so, it cannot make testable scientific predictions about specific physical systems and their specific corresponding experiences. For more on IIT, see Oizumi, M., Albantakis, L., and Tononi, G. 2014. "From the phenomenology to the mechanisms of consciousness: Integrated information theory 3.0," *PLOS Computational Biology* 10: e1003588; Hoel, E. P. 2017. "When the map is better than the territory," *Entropy* 19: 188, doi: 10.3390/e19050188.
13. See, e.g., Clarke, D. S., ed. 2004. *Panpsychism: Past and Recent Selected Readings* (New York: University of New York Press).
14. Du, S., Tao, Y., and Martinez, A. M. 2014. "Compound facial expressions of emotion," *Proceedings of the National Academy of Sciences* 111(15): E1454–E1462.
15. Goodall, J. 2011. *My Life with the Chimpanzees* (New York: Byron Preiss Visual Publications).
16. Revuz, D. 1984. *Markov Chains* (Amsterdam: North-Holland).
17. More technically, the collection of events for a measurable space is a σ-algebra, which is closed under countable union. One can generalize this to a σ-additive class, which is closed under countable disjoint union. See, e.g., Gudder, S. *Quantum Probability* (San Diego: Academic Press). One can generalize even further, to finite additive classes.
18. Revuz, D. 1984. *Markov Chains* (Amsterdam: North-Holland).
19. Hoffman, D. D., and Prakash, C. 2014. "Objects of consciousness," *Frontiers in Psychology: Perception Science*, http://dx.doi.org/10.3389/fpsyg.2014.00577.
20. Ibid.
21. Fields, C., Hoffman, D. D., Prakash, C., and Prentner, R. 2017. "Eigenforms, interfaces and holographic encoding: Toward an evolutionary account of objects and spacetime," *Constructivist Foundations* 12(3): 265–74; Fields, C., Hoffman, D. D., Prakash, C., and Singh, M. 2017. "Conscious agent networks: Formal analysis and application to cognition," *Cognitive Systems Research* 47: 186–213.

22. Kahneman, D. 2011. *Thinking, Fast and Slow* (New York: Farrar, Straus and Giroux).
23. They form the affine group AGL(4,2), and act in the geometric algebra G(4,2), the conformal spacetime algebra. Hoffman, D. D., and Prakash, C. 2014. "Objects of consciousness," *Frontiers in Psychology: Perception Science*, http://dx.doi.org/10.3389/fpsyg.2014.00577.
24. Tooby, J., Cosmides, L., and Barrett, H. C. 2003. "The second law of thermodynamics is the first law of psychology: Evolutionary developmental psychology and the theory of tandem, coordinated inheritances: Comment on Lickliter and Honeycutt (2003)," *Psychological Bulletin* 129: 858–65.
25. Faggin, F. 2015. "The nature of reality," *Atti e Memorie dell'Accademia Galileiana di Scienze, Lettere ed Arti*, Volume CXXVII (2014–2015) (Padova: Accademia Galileiana di Scienze, Lettere ed Arti). He speaks of conscious units rather than conscious agents.
26. Berkeley, G. 1710. *A Treatise Concerning the Principles of Human Knowledge.*
27. For more on the problem of demarcating science from pseudoscience, see Pigliucci, M., and Boudry, M., eds. 2013. *Philosophy of Pseudoscience: Reconsidering the Demarcation Problem* (Chicago: University of Chicago Press); Dawid, R. 2013. *String Theory and the Scientific Method* (Cambridge, UK: Cambridge University Press).
28. For instance, Mercier and Sperber (2011): "Our hypothesis is that the function of reasoning is argumentative. It is to devise and evaluate arguments intended to persuade." Tappin, van der Leer, and McKay (2017): "We observed a robust desirability bias—individuals updated their beliefs more if the evidence was consistent (vs. inconsistent) with their desired outcome. This bias was independent of whether the evidence was consistent or inconsistent with their prior beliefs . . . we found limited evidence of an independent confirmation bias in belief updating." Mercier, H., and Sperber, D. 2011. "Why do humans reason? Arguments for an argumentative theory," *Behavioral and Brain Sciences* 34: 57–111; Tappin, B. M., van der Leer, L., and McKay, R. T. 2017. "The heart trumps the head: Desirability bias in political belief revision," *Journal of Experimental Psychology: General*, doi: 10.1037/xge0000298.
29. Gould, S. J. 2002. *Rocks of Ages: Science and Religion in the Fullness of Life* (New York: Ballantine Books).
30. Dawkins, R. 1998. "When religion steps on science's turf," *Free Inquiry* 18(2): 18–19.
31. Hoffman, D. D., and Prakash, C. 2014. "Objects of consciousness," *Frontiers in Psychology: Perception Science*, http://dx.doi.org/10.3389/fpsyg.2014.00577.

Appendix: Precisely

1. Hoffman, D. D., and Prakash, C. 2014. "Objects of consciousness," *Frontiers in Psychology: Perception Science*, http://dx.doi.org/10.3389/fpsyg.2014.00577; Fields, C., Hoffman, D. D., Prakash, C., and Prentner, R. 2017. "Eigenforms, interfaces and holographic encoding: Toward an evolutionary account of objects and spacetime," *Constructivist Foundations* 12(3): 265–74; Fields, C., Hoffman, D. D., Prakash, C., and Singh, M. 2017. "Conscious agent networks: Formal analysis and application to cognition," *Cognitive Systems Research* 47: 186–213.

2. Revuz, D. 1984. *Markov Chains* (Amsterdam: North-Holland).
3. Ibid.
4. Hoffman, D. D., and Prakash, C. 2014. "Objects of consciousness," *Frontiers in Psychology: Perception Science*, http://dx.doi.org/10.3389/fpsyg.2014.00577; Fields, C., Hoffman, D. D., Prakash, C., and Prentner, R. 2017. "Eigenforms, interfaces and holographic encoding: Toward an evolutionary account of objects and spacetime," *Constructivist Foundations* 12(3): 265–74; Fields, C., Hoffman, D. D., Prakash, C., and Singh, M. 2018. "Conscious agents networks: Formal analysis and application to cognition," *Cognitive Systems Research* 47: 186–213.
5. Doran, C., and Lasenby, A. 2003. *Geometric Algebra for Physicists* (New York: Cambridge University Press), section 10.7.
6. The evolution of small-world networks is discussed, for instance, in Jarman, N., Steur, E., Trengove, C., Tyuykin, I. Y., and van Leeuewn, C. 2017. "Self-organization of small-world networks by adaptive rewiring in response to graph diffusion," *Nature Reports* 7: 13158, doi: 10.1038/s41598-017-12589-9); Newman, M. E. J. 2010. *Networks: An Introduction* (New York: Oxford University Press).

INDEX

Note: Page numbers in *italics* refer to illustrations.